NATURA

AF846405

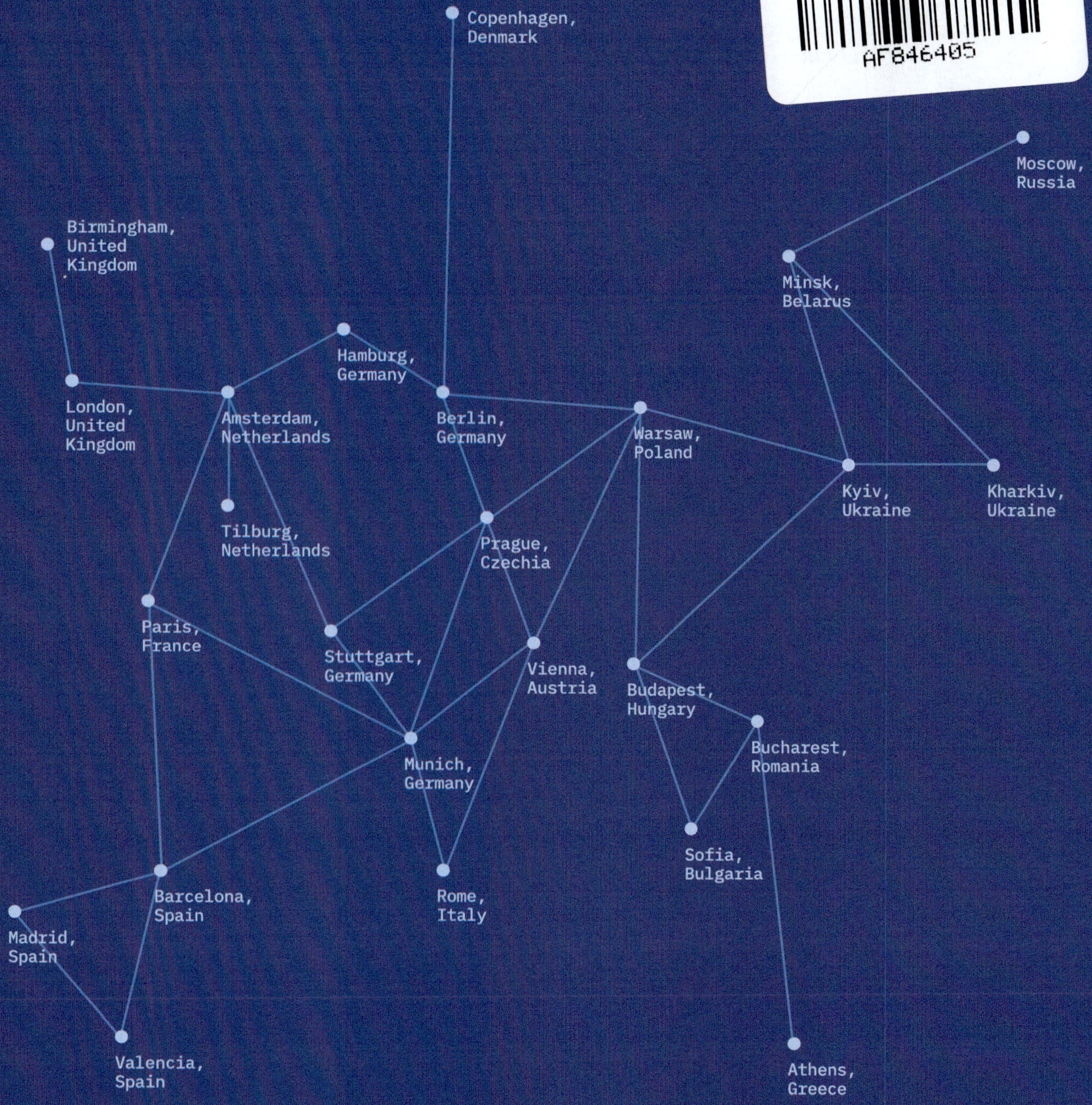

Europe Roadmap for Urban Nature-based Solutions

NYU | Global Access Publications

An imprint of
NEW YORK UNIVERSITY PRESS
New York
www.nyupress.org

This work is licensed under the Creative Commons Attribution-NonCommercial-NoDerivs 4.0 license (CC BY-NC-ND 4.0). To view a copy of the license, visit https://creativecommons.org/licenses/by-nc-nd/4.0.

© 2026 by NATURA
All rights reserved

Please contact the Library of Congress for Cataloging-in-Publication data.

ISBN: 9781479850570 (paperback)
ISBN: 9781479850624 (library ebook)
ISBN: 9781479850587 (consumer ebook)

This book is printed on acid-free paper, and its binding materials are chosen for strength and durability. We strive to use environmentally responsible suppliers and materials to the greatest extent possible in publishing our books.

The manufacturer's authorized representative in the EU for product safety is Mare Nostrum Group B.V., Doelen 72, 4831 GR Breda, The Netherlands. Email: gpsr@mare-nostrum.co.uk.

Manufactured in the United States of America
10 9 8 7 6 5 4 3 2 1
Also available as an ebook

Design & Layout by Phuong Anh Bui

FSC www.fsc.org MIX Paper | Supporting responsible forestry FSC® C013604

Lead Authors

Jarumi Kato-Huerta [1]
University of Trento

Eugenia Castellazzi [1]
University of Helsinki

Oriol Garcia-Antúnez [1]
University of Copenhagen
University of Helsinki

Davide Geneletti
University of Trento

Christopher M. Raymond
University of Helsinki

Natalie M. Gulsrud
University of Copenhagen

Contributing Authors

Filipa Grilo
University of Lisbon

Emma Campbell
Queen's University
Belfast

Loan Diep
New York University
The New School

Timon McPhearson
New York University
The New School

This report should be cited as:

Kato-Huerta, J., Castellazzi, E., Garcia-Antúnez, O., Geneletti, D., Raymond, C.M., Gulsrud, N.M., Grilo, F., Campbell, E., Diep, L. & McPhearson, T. (2025) *Europe Roadmap for Urban Nature-Based Solutions.* In: The Global Roadmap for Urban Nature-Based Solutions. New York City, NY: NATURA. https://doi.org/10.18574/nyu/9781479850624.0010001.

Contributions and Acknowledgements

Jarumi Kato-Huerta, Eugenia Castellazzi, Oriol Garcia-Antúnez, Davide Geneletti, Christopher Mark Raymond and Natalie Marie Gulsrud led data synthesis and writing of the Regional Report, with contributions from Filipa Grilo, Emma Campbell, Loan Diep, and Timon McPhearson. All authors contributed to the final recommendations and reviewed the report for accuracy and clarity. The authors thank Yaella Depietri for her inputs into the report, Thilo Wellmann, Agnieszka Sobol and Elsa Gallez for key case study inputs, Christian Albert, Mario Balzan, Erik Andersson and Chiara Cortinovis for reviewing the draft and Frédéric Lemaître for facilitating initial case study selection.

The study was supported by Natalie Pierson (research and coordination), Rebecca Tiernan, Vanya Bisht, and Dave Kendal (systematic mapping), Carole Alazki (case study data analysis), Veronica Olivotto (case study review, and copy editing), Raffaello Khan Agustin (review of reviews), and Phuong Anh Bui (design and communications). The study was also supported by Valérie Lechêne (design strategy) and Sofya Krasnaya (literature review). Timon McPhearson and Loan Diep conceptualized and led the overall Global Roadmap initiative, including regional report structure, design and final editorial review. The NATURA 'Global Roadmap for Urban Nature-based Solutions' was funded by the National Science Foundation.

[1] *These authors have contributed equally to this report and thus share first co-authorship*

Co-Leads

Timon McPhearson
NYU/The New School

Loan Diep
NYU/The New School

Advisors

Niki Frantzeskaki
University of Utrecht

Erik Andersson
University of Helsinki

Rob McDonald
The Nature Conservancy

Pippin Anderson
University of Cape Town

Harriet Bulkeley
Durham University

Yeowon Kim
Korea University

Weiqi Zhou
Chinese Academy of Sciences

Literature Review

Loan Diep
NYU/The New School

Rebecca Tiernan
Arizona State University

Natalie Pierson
NYU/The New School

Vanya Bisht
University of Waterloo

Dave Kendal
Future in Nature

Case Study Analysis

Carole Alazki
NYU/The New School

Natalie Pierson
NYU/The New School

Veronica Olivotto
NYU/The New School

Copy Editing

Veronica Olivotto
NYU/The New School

Design

Phuong Anh Bui
NYU/The New School

Funding

This report has been primarily funded by NSF (Grants #1927468 and #1927167). This project has received additional support from the CO-CARBON project, funded by the Strategic Research Council (SRC) within the Research Council of Finland (grant nos. 358254 and 358256), the SMARTer Greener Cities project through the Nordforsk Sustainable Urban Development and Smart Cities program (Project no. 95377), and the PLANNING with YOUTH project (grant no. 2019-01887) funded by FORMAS—the Swedish Research Council for Sustainable Development under the targeted call National Research Program for Sustainable Spatial Planning.

Table of Contents

Executive Summary

Urban Nature-based Solutions in Europe have gained prominence as multifunctional strategies that address intersecting challenges including climate change, biodiversity loss, and public health concerns.

The redevelopment project of the Kursaal in Abano Terme, Italy. Stefano Boeri Architetti, 2025.

Urban Nature-based Solutions (NbS) in Europe have gained prominence as multifunctional strategies that address intersecting challenges including climate change, biodiversity loss, and public health concerns. Supported by a dense policy landscape and major funding from the European Commission (EC)—Horizon Europe, LIFE, the EU Missions among others—urban NbS have evolved into key instruments for achieving climate resilience, sustainable urban development, environmental justice and social well-being. From stormwater management to urban regeneration, NbS projects are being implemented across diverse European cities.

The *Europe Roadmap for Urban Nature-based Solutions* summarizes findings on trends, challenges, and opportunities for NbS in European cities. It builds on a literature review and the combined knowledge of regional experts, presenting a synthesis of urban NbS practices in Europe illustrated by 21 urban case studies. Grounded in the latest research, the report offers actionable guidance for policy, planning, and practice. From this assessment, seven Key Messages (KM) synthesize Key Findings and Recommendations outlining the state of urban NbS in Europe and the most urgent next steps.

KEY MESSAGE 1:

Europe is strongly investing in urban NbS, reflecting a vision to position itself as a global NbS leader. Europe's investments into NbS reflect its ambition to lead globally. Through strategies such as the EU Biodiversity Strategy and the Green Deal, Europe aims to lead by example. A cornerstone of this vision is the European Commission's own definition of NbS as "solutions that are inspired and supported by nature, which are cost-effective, simultaneously provide environmental, social and economic benefits and help build resilience." Yet translating ambition into widespread implementation remains an ongoing challenge in many sub-regions.

Key findings: Europe's experience with NbS builds on a long legacy of urban greening efforts. As of 2025, 887 million EUR (over 1 billion USD) have been invested into NbS Research and Innovation by the European Commission since 2014 (European Commission, 2025). These efforts, combined with growing scientific evidence and long-standing municipal policies, have positioned Europe as a pioneer in mainstreaming urban NbS globally.

Key recommendations: Solidify Europe's NbS leadership by deepening investment in implementation at the local level and fostering mutual learning among cities, including outside the European Union. Sharing best practices across contexts can improve effectiveness while reinforcing Europe's capacity to lead by example within the region and globally.

KEY MESSAGE 2:

Urban NbS initiatives in Europe have focused on the restoration and creation of new green spaces, with not enough efforts dedicated to safeguarding and managing existing ecosystems in urban areas. Most urban NbS in Europe are based on restoring or creating new green spaces or modifying existing ones, while conservation and preservation of natural areas are less common.

Key findings: The dominance of what is known under the EC's classification as "Type II (restoration of ecosystems) and III NbS" (creation/modification) reflects urban densification pressures and the compact city model. However, the limited integration of conservation-based NbS (Type I) reduces opportunities for preserving remnant biodiversity and landscape connectivity. The EU Restoration Law creates potential for improving conservation-based approaches by setting legally binding targets to increase tree canopy coverage, including within cities.

Key recommendations: Embed biodiversity as a core objective across all urban NbS, prioritizing efforts that recognize the value of existing urban ecosystems that enhance ecological integrity. This should be supported by monitoring strategies that measure progress made against biodiversity targets, including those set by the EU Restoration Law, with accurate data to capture ecological processes and species interactions.

KEY MESSAGE 3:

Multi-stakeholder partnerships are central to project success, though private sector engagement remains weak. Collaborative governance models involving municipalities, academia, and NGOs have now become more common, yet private sector participation is inconsistent.

Key findings: EU funding schemes such as Horizon 2020 and Horizon Europe have catalyzed collaboration and prompted new funding models, complemented by independent initiatives from various government levels. Despite the advances achieved when engaging the private sector (e.g., through the nature-positive economy effort), private corporations continue to perceive NbS investment as risky. This is a gap noted by the IPBES' Transformative Change Assessment, which states that private finance accounts for only around 17 percent of total NbS investment.

Organizations like the European Investment Bank are seeking to address this gap through dedicated de-risking instruments and blended finance approaches. The need for longer term funding and involvement of the private sector underscores the importance of multi-actor collaboration and innovative funding for scaling NbS.

Key recommendations: Develop blended finance mechanisms and incentives to increase private sector investment, including risk-sharing models to help scale up and support NbS over the long-term. Strategic frameworks that embed economic sustainability (such as those combining spatial planning with early-stage enterprise support) could help cities explore self-financing pathways for NbS. For instance, initiatives that leverage opportunity mapping, nature-based business models, and partnerships to unlock public and private funding are demonstrating how governance innovation and open-source tools can support long-term investment, including through emerging carbon and ecosystem service markets.

KEY MESSAGE 4:

Co-creation and inclusion are emerging as powerful enablers of NbS, but require deeper institutional support, recognition of value pluralism, and sustained community stewardship. While co-creation can enhance procedural justice and local ownership, EU-funded frameworks often promote standardized approaches to NbS implementation that risks overlooking meaningful place-based values – particularly in regions with distinct socio-political histories or cultural views of nature.

Key findings: Participatory approaches to engage citizens in planning, implementation and monitoring are gaining traction, yet approaches often reflect a narrow, instrumental framing of nature. Greater recognition of the diverse values that people attribute to urban nature is needed, particularly those grounded in cultural, and relational perspectives.

Key recommendations: Design participatory processes that explicitly include diverse knowledge systems and value perspectives. Encourage municipalities to move beyond instrumental justifications for NbS and adopt more inclusive framings that reflect local values and worldviews. Support stewardship-based funding models that empower communities to sustain NbS over time. Local governments can strengthen ownership and long-term care by channeling resources through innovative mechanisms, such as innovation vouchers, that enable residents and local actors to co-develop and maintain NbS.

KEY MESSAGE 5:

Europe has a robust policy and regulatory framework for NbS, but further efforts are needed for policy coherence and institutional coordination. Regulatory and policy enablers of NbS exist across international, regional, and local levels. Some policies integrate multiple governance scales, while others focus on specific levels, such as municipal greening strategies or national restoration plans aligned with EU goals. Strengthening these connections will enhance NbS effectiveness.

Key findings: While frameworks like the EU Restoration Law can support urban NbS, local implementation often suffers from fragmented responsibilities, limited enforcement, and unclear regulatory pathways. Misalignment between governance levels and policy silos hinders coherent implementation. Uneven implementation can be observed across sub-regions, with more limited implementation in parts of Eastern and Southeastern Europe for example.

Key recommendations: Promote policy coherence and institutional coordination across governance levels that reduce siloed decision-making and foster collaboration across departments and stakeholders. Embed NbS targets in local plans and ensure regulatory clarity on procurement, land use, and performance standards, while aligning city and national actions with EU restoration and other European biodiversity goals to safeguard political and financial support.

KEY MESSAGE 6:

Technical gaps – notably in long-term monitoring and evaluation – undermine the sustainability and resilience of NbS. Many NbS are under pressure from climate extremes, limited maintenance capacity, and changing urban contexts, yet evidence on their long-term resilience is limited. These gaps are reinforced by unclear standards, fragmented responsibilities for upkeep, and short political funding cycles that favor quick wins over longitudinal learning.

Key findings: Droughts, such as those in 2018, have stressed urban trees and green infrastructure, revealing vulnerabilities in NbS design and management. Few projects systematically monitor degradation over time, and there is little data on how co-benefits evolve as NbS age. Compound events (e.g., heatwaves plus water restrictions) further expose these weaknesses, and without baselines and consistent indicators it is hard to distinguish normal aging from avoidable design or management failures.

Key recommendations: Expand research and funding for long-term monitoring of NbS performance and adaptive management. Cities should plan for changing functionality over time by designing NbS with built-in resilience to climate stressors. Set common indicators and minimum monitoring horizons (e.g., 5–10 years) tied to funding contracts, and require open data reporting so results are comparable and lessons can be scaled across cities.

KEY MESSAGE 7:

Scaling remains a key challenge despite successful local pilots. Scaling NbS to meet the full magnitude of urban challenges requires a better match between the size of interventions and the scope of environmental needs – particularly in Europe's densely populated and highly planned urban landscapes. NbS often fall short of addressing the full scale of environmental and social challenges in cities, due to a lack of mechanisms for expanding reach.

Key findings: While successful pilots have scaled vertically (institutionally) and spatially (across sites) (e.g., Lisbon's allotment garden network), most projects do not undergo such transitions. There is limited evidence of NbS being designed to match the scale of identified urban risks or ecological deficits.

Key recommendations: Develop needs-based scaling strategies that assess the scope of urban challenges and link them to appropriate NbS solutions. Foster horizontal, vertical, and temporal scaling by embedding NbS into city-wide planning instruments and aligning them with long-term development goals.

List of Figures

List of Boxes

List of Tables

Acronyms and Abbreviations

Acronyms and Abbreviations	Full Term/Description
BCC	Belfast City Council
BCR	Brussels-Capital Region
CAL	Clever Action Labs
CEIS ARTE	Reception Center for Asylum Seekers in Finale Emilia
DKK	Danish Krone
EIDUS	Integrated Strategy for Sustainable Urban Development
ELYkeskus	Centre for Economic Development, Transport and the Environment
ELP	Emscher Landscape Park
EC	European Commission
EbA	Ecosystem-based Adaptation
ERDF	European Regional Development Fund
ES	Ecosystem Services
ESD	Education for Sustainable Development
EU	European Union
EUR	Euro
FEUT	Fundación Ecología Urbana y Territorial (Urban and Territorial Ecology Foundation)
GBP	British Pound Sterling
GEMAPI	Gestion des Milieux Aquatiques et Prévention des Inondations (Management of Aquatic Environments and Flood Prevention)
GI	Green Infrastructure
GIS	Geographic Information Systems
HOFOR	Hovedstadsområdets Forsyningsselskab (Greater Copenhagen Utility)
IBA	International Building Exhibition Emscher Park

Acronyms and Abbreviations	Full Term/Description
IPBES	Intergovernmental Science-Policy Platform on Biodiversity and Ecosystem Services
IUCN	International Union for Conservation of Nature
LAC	Latin America and the Caribbean
LHE	Large Housing Estates
MCDA	Multi-Criteria Decision Analysis
MENA	Middle East and North Africa
MITECO	Ministry for the Ecological Transition and Demographic Challenge
NATURA	Nature-based Solutions for Urban Resilience in the Anthropocene
NATURVATION	Nature-based Urban Innovation
NbS	Nature-based Solutions
NbCSS	Nature-based Climate School Shelters
NGO	Non-Governmental Organization
NRE	Nutsbedrijven Regio Eindhoven (Utilities in the Eindhoven Region)
PPP	Public-Private Partnership
PPGIS	Public Participatory Geographic Information Systems
PTM	Piano Territoriale Metropolitano (Metropolitan Territorial Plan)
QUB	Queen's University Belfast
RFI	Rete Ferroviaria Italiana (Italian Railway Network)
RUNTS	National Third Sector Register
SafN	Solutions d'Adaptation Fondées sur la Nature (Nature-based Solutions)
SDG	Sustainable Development Goal
SME	Small and Medium-sized Enterprise
SUDS	Sustainable Urban Drainage Systems
TCV Scotland	Conservation Volunteers Scotland
UN	United Nations
UN-HABITAT	United Nations Human Settlements Programme
UNaLab	Urban Nature Labs
URBiNAT	Urban Innovative & Inclusive Nature
USD	US Dollar
Uf-NbS	Urban Forestry and Nature-based Solutions
WSUD	Water Sensitive Urban Design
WWF	World Wide Fund for Nature

1 | Introduction

Nature-based Solutions are increasingly recognized worldwide and in Europe as critical solutions for addressing multiple societal challenges simultaneously, ranging from climate risks to biodiversity loss, human health and well-being.

A Green Promenade for Cagliari, Italy. Stefano Boeri Architetti, 2025.

1.1 Urban Nature-based Solutions in Europe

Nature-based solutions (NbS) are increasingly recognized worldwide and in Europe as critical solutions for addressing multiple societal challenges simultaneously, ranging from climate risks, to biodiversity loss, human health and well-being. Interest in urban NbS in Europe began over a decade ago, gaining momentum in 2014 when the European Commission formed an expert group to explore ways of reintegrating nature into urban areas (Faivre et al., 2017). An analysis of funding allocation data from the EU NbS Project Database (H2020, FP7, LIFE, BiodivERsA, and Interreg) shows that funding for NbS projects in European programmes increased significantly between 2011 and 2017, rising from under 25 million EUR (34.7 million USD) to over 100 million EUR (113 million USD) annually. Between 2017 and 2020, funding levels stabilized at just above 100 million EUR per year (EL Harrak and Lemaitre, 2023). Outside the European Union, NbS also gained traction as one of the main modern urban planning strategies (Dushkova et al., 2024). Across European cities, NbS have been implemented for many different reasons, including to provide multiple benefits for people and biodiversity, supporting the delivery of a diverse range of ecosystem services, from urban cooling and flood control benefits to water quality, public health, education, recreation, and more. For NbS to be effective, research and practice consistently emphasize the need for their adaptability to local social, ecological, geographic, economic, and political contexts (Albert et al., 2019; Andersson et al., 2019, 2021; Schröter et al., 2022).

Europe presents a complex mosaic of nations and historical regions, shaped by centuries of changing borders, alliances, and conflicts. While the European Union (EU) is the most influential transnational organization in the region, comprising 27 Member States with some benefiting from free movement of people and a shared currency (the Euro), Europe as a whole includes more countries—44 are considered in this study—integrated into various political and economic frameworks.

Europe is a region rich in biodiversity, encompassing diverse bio-climatic zones, landscapes, and geological features, each facing both distinct and overlapping social-ecological challenges. The continent's diverse ecosystems range from rugged coastlines to towering mountain ranges, supporting an array of biodiversity. Western Europe is characterized by temperate climates, Mediterranean climates dominate the south, continental climates prevail in the east, and polar climates in the far north. Despite this natural wealth, Europe's ecosystems are under growing pressure from climate change, a challenge especially acute in urban areas.

Currently, 75 percent of Europe's population lives in cities, which is projected to rise to 85 percent by the end of the century (United Nations, 2018). Between 2018 and 2020, urban expansion occurred largely at the expense of heathlands and shrublands, contributing to land degradation and habitat loss (Maes et al., 2020). This trend, coupled with climate change, habitat fragmentation, and other human activities, intensifies the threats to local and regional biodiversity. Urban population growth also drives greater densification in cities, leading to significant variation in the quantity and quality of urban green and blue spaces. This highlights the potential of – and urgent need to – adopting and deploying urban NbS to enhance urban resilience and counter biodiversity loss.

Currently, 75 percent of Europe's population lives in cities, which is projected to rise to 85 percent by the end of the century (UN, 2018). This highlights the potential of—and urgent need to—adopting and deploying urban NbS to enhance urban resilience and counter biodiversity loss.

Social-ecological Challenges in European Urban Areas

European urban areas face complex social-ecological challenges shaped by diverse climates, urban growth patterns, and environmental pressures. Cities in Europe are particularly vulnerable to the impacts of climate change. These include rising temperatures and more frequent extreme weather events, such as cloudburst precipitations, flooding (e.g., DANA in Valencia, Spain, October 2024) (Figure 3), storm surges (e.g., storm Bodil in Denmark, Dec 2013) and longer heat waves (e.g., heatwave in Southern Europe of July 2023). Dense infrastructure, impervious surfaces, and high population densities amplify these challenges, creating urban heat islands and increasing the risk of flooding.

DANA in Valencia, Spain, October 2024. Manuel Pérez Garcia and Estefania Monerri Mínguez, 2024.

Urban growth has further complicated these issues. Expanding populations, driven largely by net migration and immigration, have increased densification in many cities, shifting trends from earlier periods of declining population density (e.g., 2006–2012) to increasing density between 2012 and 2018 (Cortinovis, Geneletti and Haase, 2022). This densification pressures critical urban services and has contributed to unsustainable land-use practices, such as increased soil sealing and the loss of peri-urban agroecosystems. These trends undermine ecological balance and exacerbate vulnerabilities to climate change. Urban regeneration is another pressing concern. Many neighborhoods experience economic and social decline, requiring opportunities for revitalization by improving public spaces and fostering community well-being (Zwierzchowska et al., 2019). Moreover, urban densification often reduces green spaces per capita, often falling below global recommendations (Kabisch et al., 2016; Wolff and Haase, 2019) and making it harder to provide adequate recreational areas for residents. This scarcity can negatively impact public health, as green spaces are essential for physical activity, mental well-being, and social cohesion. Therefore, integrating green infrastructure into urban planning is vital to ensure that as cities grow denser, they remain livable and resilient, offering residents the benefits of nature amidst urban landscapes.

Storm Bodil in Denmark, December 2013. julochka, 2013.

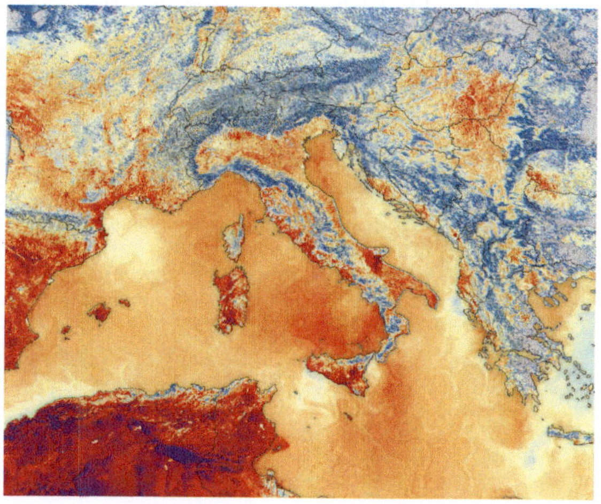

Heatwave in Southern Europe, July 2023. The European Space Agency, 2025.

These trends have informed the creation of the EU's Urban Agenda, which defines 14 priority areas below. Central to this framework are the EU Climate Law, which establishes legally binding targets to achieve climate neutrality by 2050, and the EU Biodiversity Strategy, which aims to restore ecosystems, protect natural habitats, and promote sustainable urban development. A recent and globally pioneering example of the EU's growing emphasis on biodiversity is the Nature Restoration Regulation, which obliges member states to restore degraded ecosystems—including those in urban areas. Together, these initiatives highlight the critical role of NbS as strategic approaches to integrate nature's structure and processes into urban settings, thereby addressing environmental challenges while supporting broader EU climate and biodiversity goals (European Commission, 2024).

1.2 The State of Knowledge on Urban Nature-based Solutions in Europe

NbS in Europe are now a cornerstone of efforts to tackle the intersectional challenges of climate change, public health, and biodiversity loss, leading to numerous initiatives to test and refine their applications. Europe has also emerged as a global leader in urban NbS research. Europe and Asia account for the highest volume of empirical, place-based research on urban NbS through studies that document real-world interventions, projects, or case studies (Figure 1). This regional pattern reflects where urban NbS have been most actively studied in practice over time, while highlighting persistent geographic imbalances: regions such as Latin America & the Caribbean (LAC), Oceania, Africa, and the Middle East & North Africa (MENA) remain comparatively underrepresented in the literature.

Europe's research leadership is further emphasized in Figure 2, which shows the proportion of Europe-affiliated authors contributing to urban NbS publications across different world regions. European researchers dominate the literature within their own region and maintain a strong presence in studies focused on other parts of the world, including LAC, Africa, Oceania, and MENA. While their representation is relatively lower in publications focused on the US & Canada and Asia, European-based authors remain more consistently represented across all global

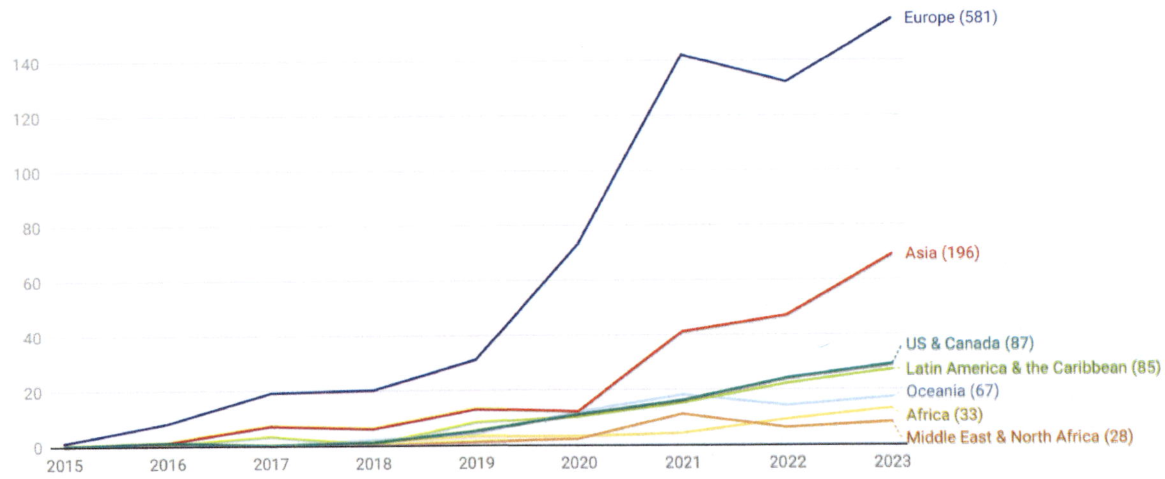

Figure 1: Regional Distribution of Empirical Academic Literature on Urban Nature-based Solutions (2015–2023).

regions than those from any other region. This illustrates Europe's outsized role not only in producing urban NbS research, but also in shaping the global knowledge discourse.

However, despite this strong academic foundation, mainstreaming the NbS concept in practice continues to face significant barriers. These include a lack of policy integration across administrative silos, limited technical expertise and standardization compared to grey infrastructure, and insufficient engagement of local communities in NbS planning and implementation.

Previous assessments and databases have significantly contributed to the research and implementation of NbS across Europe, notably to better understand those challenges and propose ways to overcome them. For instance, the BiodivERsA network, funded by the EU, has advanced NbS research by coordinating and supporting biodiversity research programs across Europe, producing guidelines on NbS best practices, and assessing the contributions of NbS to biodiversity and climate resilience (Akoumianaki and Pakeman, 2023). Similarly, the EKLIPSE Expert Working Group has put together assessment frameworks for evaluating and supporting the planning of NbS (Raymond, Breil, et al., 2017). NetworkNature's own NbS knowledge assessment has supported the making of an NbS Roadmap for the EU (EL Harrak and Lemaitre, 2023), notably to bridge the gap between NbS research and implementation and to integrate NbS research and innovation into policy. The Oppla information hub (https://oppla.eu/) is an online open-access repository of EU NbS projects and toolkits. Through its atlas, The NATURVATION

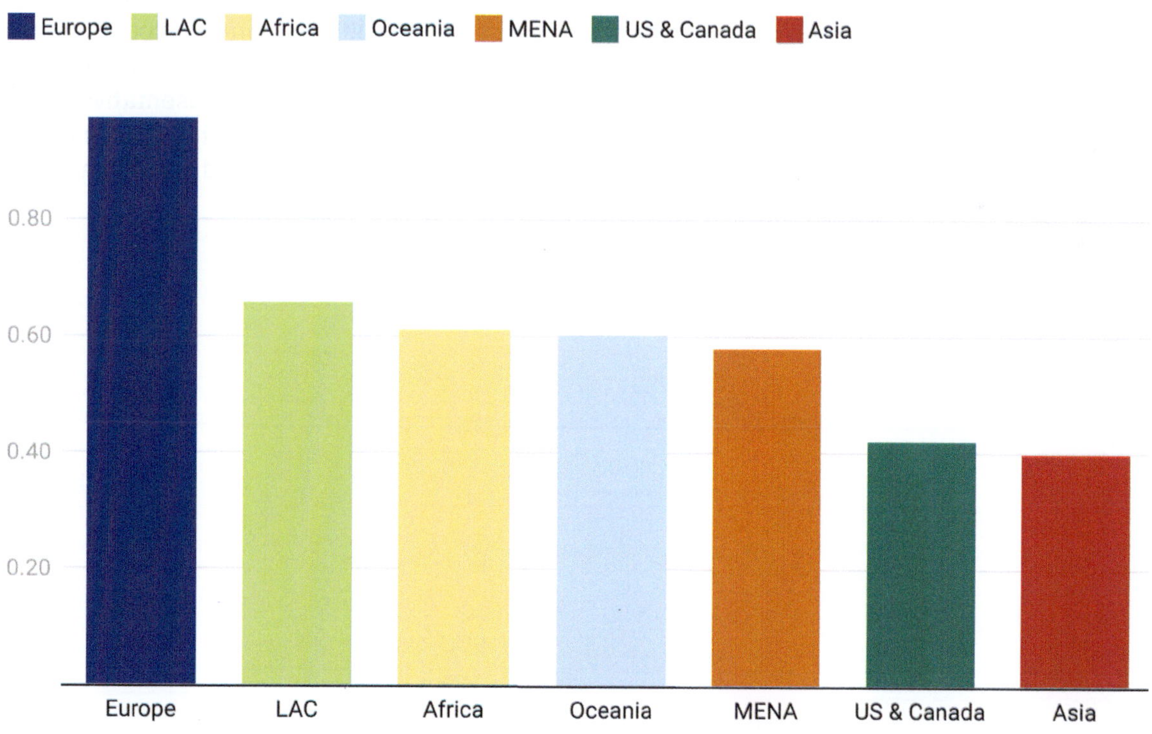

Figure 2: Proportion of Europe-Affiliated Authors Contributing to the Urban Nature-based Solutions Literature Across Regions.

Note: The proportions shown do not add up to 1, as papers may contribute to multiple regions through either author affiliation or study focus. Each paper can be counted once per region based on the diversity of author affiliations. For instance, a paper with authors affiliated in Europe (3), US & Canada (2), and LAC (1) would contribute one count to each of those regions. Additionally, if a paper addresses multiple regional case studies (e.g., Africa and Europe) and includes author affiliations from Africa (1) and MENA (2), those affiliations would each be counted once under both regions. All author affiliations were counted, including when an author had multiple affiliations.

2017-2022 project documented over 1,000 NbS cases across Europe and beyond, leading to the most comprehensive database of urban NbS projects to date. We build on these approaches to provide an updated regional assessment of the state of knowledge and practice of NbS in Europe.

1.3 Contributions of This Report

The *European Regional Roadmap for Urban Nature-based Solutions* is one of seven regional assessment reports in the *NATURA Global Roadmap for Urban Nature-Based Solutions*. This effort builds on previous assessments and databases, specifically focusing on urban contexts. The report argues that despite the wealth of knowledge in the region, many knowledge gaps remain in the urban context, which limits the ability of urban policymakers, planners and city dwellers to implement NbS effectively. The urban focus of this NbS assessment is particularly relevant for the region, as European cities are increasingly leveraging NbS to integrate biodiversity into urban spaces, delivering essential ecosystem services and enhancing the quality of life for residents. We provide key findings from this regional contribution to the global assessment process. Advancing from previous assessments and databases mentioned in 1.2, the primary contribution of this report lies in its mixed and empirically grounded methodology. This report integrates insights from an extensive review of academic and grey literature, and the collective expertise of a multidisciplinary expert group. Furthermore, it is enriched by granular data derived from 21 regionally representative case studies, offering context-specific examples of how NbS are implemented and operate across diverse urban settings.

2 | Scope and Process

The Ca'delle Alzaie residential project in Treviso, Italy. Stefano Boeri Architetti, 2021.

2.1 Overall Approach

This report is part of the *NATURA Global Roadmap for Urban Nature-based Solutions*—a coordinated initiative to synthesize the state of knowledge and practice across seven world regions (US-Canada, Latin America & the Caribbean, Europe, the Middle East & North Africa, Africa, Asia, Oceania). The methodology follows a framework used across the global and other regional assessments, enabling both cross-regional comparison and region-specific analysis.

Within this shared framework, the European report offers a comprehensive perspective on NbS in European cities, drawing on two main sources of knowledge: a review of evidence from research on urban NbS and a set of case studies of practice that detail the operationalization of urban NbS. This dual perspective is further enriched by the expert knowledge of the authorship team, which not only validates the process but also provides additional insights into the European urban NbS landscape. The aims and objectives of this report are threefold: (1) to explore patterns and trends in NbS research across the region, (2) to highlight case studies that demonstrate how NbS are being integrated into policy, planning, and practical implementation, and (3) to synthesize existing knowledge, identify key knowledge gaps, and highlight opportunities for scaling and adapting NbS to different contexts. This process was carried out through an iterative, collaborative and consensus-oriented co-production approach, drawing on regional expertise and global coordination to ensure both comparability and local relevance (Figure 3).

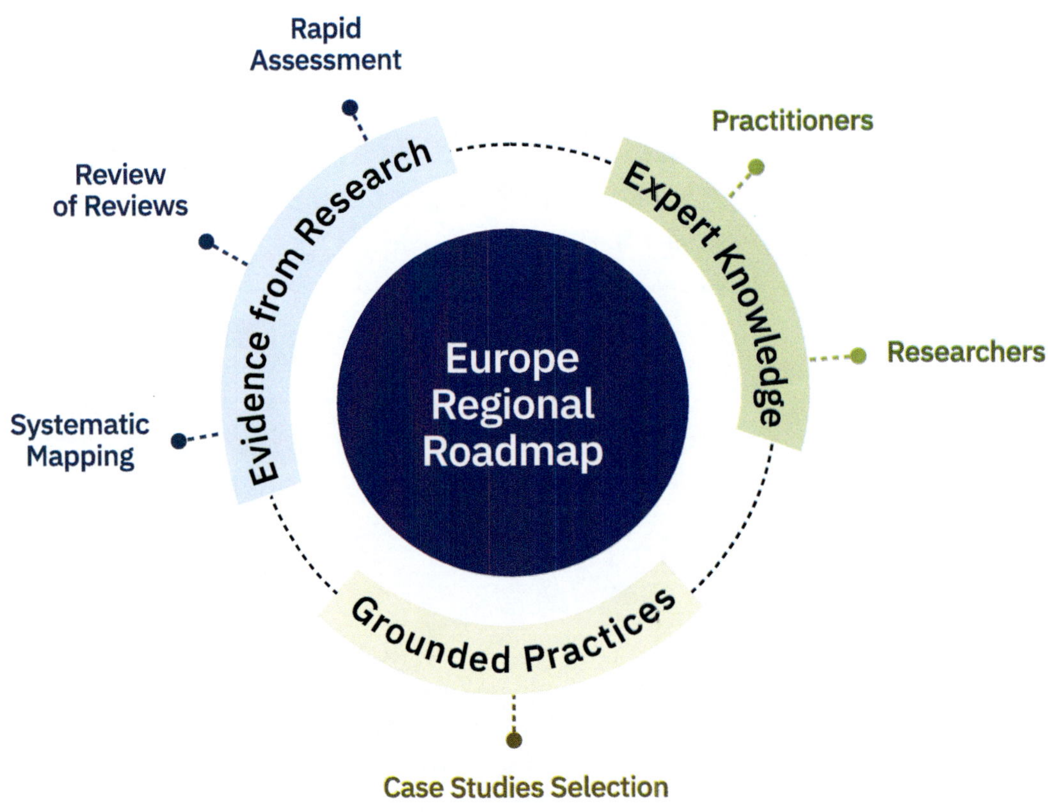

Figure 3: Integrative approach to the NATURA Europe Roadmap for Urban Nature-based Solutions.

2.2 Methods to Assess the State of Research

To assess trends and gaps in current urban NbS research, the team applied a three-part methodology designed to capture both broad patterns and detailed insights across the field. These methods formed part of a coordinated approach used in regional assessments for the NATURA Global Roadmap, enabling comparative analysis across regions and the global literature. This approach synthesized publication trends, thematic focus, and geographic distribution in the peer-reviewed literature. Grey literature was also included selectively to fill gaps where peer-reviewed material was scarce. Literature reviews included a rapid assessment, a review of reviews and a systematic mapping (see Appendix 7.1 for details)

Rapid Assessment

The team conducted a broad, high-level scan of English-language academic literature on urban NbS and related concepts (e.g., Ecosystem Services, Green Infrastructure, etc.), mapping when and where NbS appeared in comparison to similar terms and highlighting regional hotspots and publication trends over time. A systematic search was carried out in SCOPUS, and bibliometric analyses were performed to examine the evolution and distribution of the literature.

Review of Reviews

A targeted review was undertaken to synthesize trends, knowledge gaps, and recurring challenges as reported in existing review articles on urban NbS. Using defined search queries, review articles were identified and screened for relevance to urban contexts, NbS-related concepts, and systematic or scoping methodologies. Data extraction was guided by a structured matrix, adapted from prior work, to capture key information on NbS concepts, review types, thematic and geographic trends, and gaps across ecological, social, and governance dimensions.

Systematic Mapping

This most granular phase focused on primary studies explicitly engaging with the term "Nature-based Solutions" in urban contexts. Only empirical, place-based articles were retained following a rigorous screening process, resulting in a curated dataset. Geographical, bibliometric, and thematic trends were analyzed using RStudio and VOSViewer, with content analysis based on a validated text-mining approach to assess how societal challenges and ecosystem types are addressed across regions.

2.3 Methods to Assess the State of Practice

The review of practice aimed to build a regional profile of urban NbS implementation through a set of representative case studies, selected via an expert-based process. The goal was not to compile an exhaustive inventory, but to reflect the diversity of NbS types, geographies and implementation contexts across Europe. The selection process followed four main steps (Figure 4).

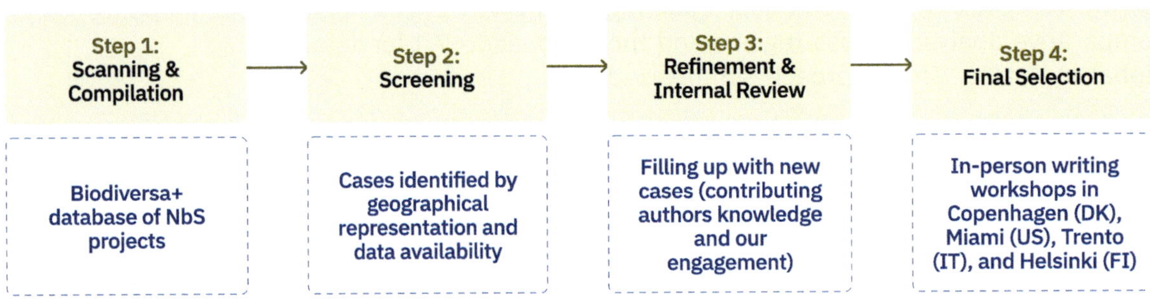

Figure 4: Steps of the Case Study Selection Methodology.

STEP 1

Scanning & Compilation

The initial pool of cases was drawn from BiodivERsA-funded projects (2015-2021), prioritized for their robust and accessible data. To assess the representativeness of these cases, the pool was cross-checked against the NATURVATION project's Urban Nature Atlas, which catalogues NbS in 100 European cities (See Appendix A for details). Additional sources included grey and academic literature as well as cases identified through the professional networks of the European Node and contributing authors.

STEP 2

Screening

A set of agreed criteria guided the screening process:

- **Representation:** Ensuring diversity of geographical locations, NbS types (e.g. parks, gardens, green roofs) and societal challenges addressed (e.g., water management, regeneration, land-use, urban development).
- **Availability of information:** Prioritizing cases with sufficient, verifiable documentation to support thorough analysis, including details on drivers, challenges addressed, NbS types, intervention scales, financing arrangements, institutional instruments, actors, ecosystem services provided, and implementation challenges.
- **Collaboration and previous knowledge:** Given preference to cases where European Node members had direct knowledge or experience, enabling deeper contextual understanding.

STEP 3

Refinement & Internal Review

The pool was refined through iterative internal discussions to balance representativeness and thematic coverage. Identified gaps—particularly geographic and sectoral—were addressed by supplementing the initial selection with new case studies sourced through extended networks. This ensured broader coverage across Europe's varied socio-ecological and governance contexts.

STEP 4

Final Selection

The final shortlist of 21 case studies was confirmed through collaborative synthesis workshops held in Copenhagen (DK), Miami (USA), Trento (IT), and Helsinki (FI). These workshops brought together contributing authors to synthesize and finalize findings into a coherent narrative for the roadmap.

The selected case studies represent a geographically diverse sample, spanning Northern, Western, Central, and Southern Europe (Figure 5). However, fewer case studies were available from Eastern and Southeastern Europe and the Baltic region. These gaps reflect limited accessible and language barriers that constrain access to relevant information. Addressing these challenges in future research will be critical to develop a more comprehensive understanding of NbS implementation across Europe's diverse urban contexts.

Figure 5: Europe's Regional Delineation Used in This Report. The boundaries and names shown, and the designations used on the map do not imply official endorsement or acceptance by NATURA

2.4 Team and Positionality

This report is the product of a collaborative effort by a core team of eight Europe-based researchers from diverse disciplinary backgrounds, including natural resource management and governance, urban planning, ecology, and sustainability science. While our geographic base inevitably informs our perspective – particularly through exposure to European policy frameworks and cultural assumptions, for example on human-nature relationships – we have actively worked to expand the scope of our understanding through disciplinary dialogue and engagement with global knowledge.

We acknowledge the limitations of our team's composition, including the absence of direct representation from Eastern European contexts. To help address this, we have incorporated case studies and literature reflecting region-specific practices. Recognizing that academic work can sometimes overlook practice-based insights, we also sought to reflect practitioner knowledge by drawing from grey literature and field-based reports.

The relevance of the EU and the EC in the transnational environmental policy sphere must be carefully acknowledged. This leadership also needs to critically reflect on Europe's historical context and current global influence, including the legacy of post-colonial policy dynamics and the transfer of policy ideas within and beyond European borders. The NATURA European Roadmap aims to contribute to a more inclusive, pluralistic, and equitable framing of NbS. This process begins by acknowledging the inherent biases in the report's creation and highlights the critical role of external collaborations and knowledge-sharing.

Natalie Gulsrud and Eugenia Castellazzi having a discussion at the NATURA All-Hands Meeting in Miami. Loan Diep, 2024.

3 State of Urban Nature-based Solutions Research and Practice in Europe

The Wonderwoods vertical forest project in Utrecht, the Netherlands. Stefano Boeri Architetti, 2025.

3.1 Trends

3.1.1 Adoption of Nature-based Solutions and Related Concepts in the Region

KEY FINDING 1:

Europe's leadership in NbS is rooted in a long tradition of urban greening, reinforced by the strategic definition, substantial funding and promotion of NbS as a unifying concept with global reach. However, publication trends show that terms like ecosystem services and green infrastructure still dominate European literature.

Europe's leadership in NbS is grounded in a longstanding tradition of urban greening that predates the concept itself. European cities have historically implemented a wide range of practices to manage water, heat, biodiversity, and urban livability – often without referring to these efforts as "nature-based solutions." With the emergence of the NbS concept in the 2010s, largely propelled by EU institutions, the European Commission took a

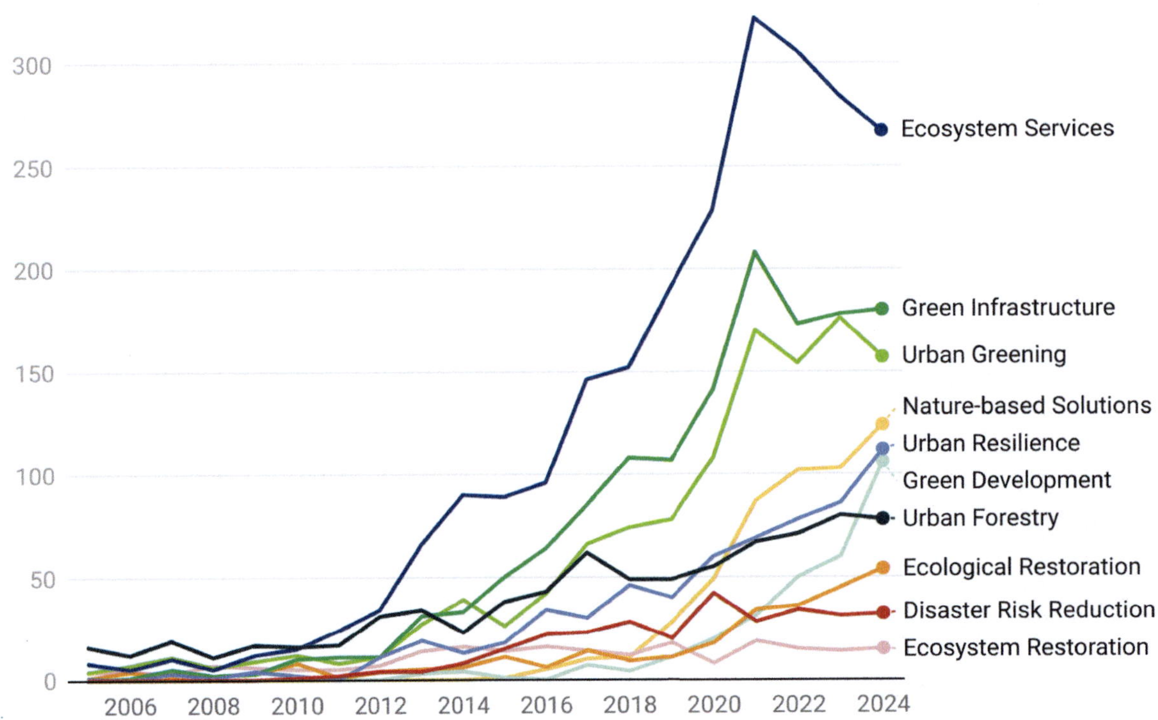

Figure 6: Trends in Top 10 NbS-Related Concepts in Urban Academic Literature: Europe (2005–2024). Note: Only papers that include a selected concept that appear in the tile, abstract, or author keywords of an article appear here.

proactive role in defining and mainstreaming NbS as an umbrella concept. The European Commission has invested over €100 million annually in NbS-related programs since 2017, supporting thousands of projects under Horizon 2020 and other frameworks (European Commission, 2025). These efforts – alongside an increasingly rich scientific knowledge base and the active involvement of municipalities – have made Europe a global reference for urban NbS development.

Given the EU's central role in conceptualizing and promoting NbS, the European Commission's definition has been widely adopted across the region. This definition frames NbS as actions "inspired and supported by nature, which are cost-effective, simultaneously provide environmental, social, and economic benefits and help build resilience" (European Commission, 2015). In practice, the concept encompasses a broad range of related strategies, including restorative solutions (e.g., ecological and landscape restoration and rehabilitation), issue-specific strategies (e.g., ecosystem-based adaptation and disaster risk reduction), infrastructure-focused (e.g., green infrastructure development), management practices (e.g., integrated coastal zone and water resource management) and protection (e.g., area-based management and conservation) (Cohen-Shacham et al., 2019).

This diversity of approaches is reflected in the academic literature. The rise of peer-reviewed publications for NbS-related concepts from 2005-2024 shows that ecosystem services, green infrastructure, urban greening, and only more recently nature-based solutions have dominated the field, while others like disaster risk reduction and ecosystem restoration are emerging but remain less prevalent (Figure 6). Additional terms like water sensitive urban design (WSUD), low impact development, sustainable urban drainage systems (SUDS) and ecological engineering (European Commission, 2023) are also used to reflect specific disciplinary or sectoral approaches. While many European cities adopt the EC definition directly, others tailor the framing to address particular local challenges or socio-environmental goals. In some cases, innovative concepts like URBiNAT's "healthy corridors" show how NbS are being adapted through participatory design and culturally grounded approaches.

3.1.2 | Types of Nature-based Solutions Projects

KEY FINDING 2:

The influence of compact city planning and urban densification drives the predominance of NbS focused on creating new green elements or modifying existing ecosystems. The limited emphasis on ecological conservation and restoration in urban contexts limits opportunities to preserve remnant biodiversity and enhance ecological connectivity.

Nature-based Solutions Projects Typologies

The European Commissions defines NbS as an umbrella term that includes different actions to sustain and maintain biodiversity and enhance human well-being. One of the most common ways to classify NbS in Europe is based on the degree of human intervention and ecosystem modification, typically distinguishing three main types – though the boundaries between them are highly fluid (Eggermont et al., 2015). In urban contexts, however, the predominance of NbS focused on creating new green elements or modifying existing ecosystems reflects the influence of compact city planning and urban densification (Evers, Katurić and Van Der Wouden, 2024). For instance, urban NbS often take the form of new or transformed green and blue infrastructure – such as green roofs, rain gardens, and bioswales – which aim to maximize multifunctional ecosystem services within constrained city spaces.

When it comes to the type of ecosystems where NbS projects are implemented, we can see that, according to empirical urban nature-based solution studies, Europe mirrors the global trend (Figure 7). Green ecosystems (primarily vegetated areas such as parks, forests, gardens, and green roofs) are the most prominent in the literature, making up over half of the studies that specify an ecosystem type. Blue ecosystems, including lakes, rivers, and marine environments, as well as Blue-Green ecosystems, which integrate transition zones dominated by water and vegetation (e.g., wetlands, riparian areas), are also represented

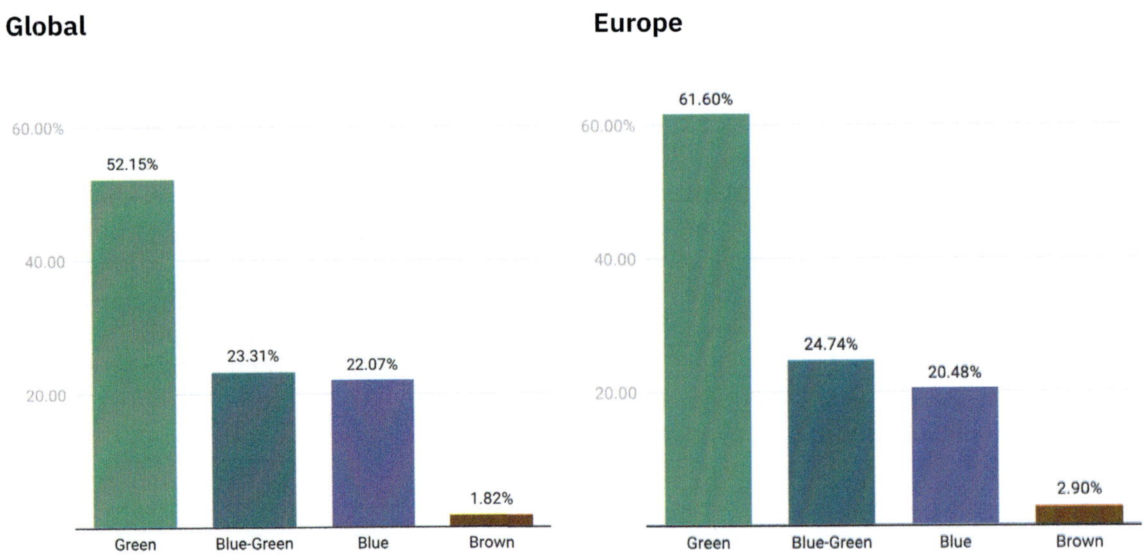

Figure 7: Proportion of Ecosystem Types Addressed in Empirical Urban Nature-based Solutions Studies: Global vs. Europe. Note: Only studies that mention an ecosystem type in the title, abstract, or authors keywords are included in this figure. Proportions shown do not add up to 1, as one paper may include multiple ecosystem types.

Implementation of an extensive green roof on top of the new IKEA building in the fast-developing neighborhood of Vestebro, Copenhagen, Denmark. Oriol Garcia-Antúnez, 2025.

but to a lesser extent. Brown ecosystems, such as brownfields and post-industrial or derelict sites undergoing ecological regeneration, are rarely studied in comparison. In the Europe-focused subset, Green ecosystems are even more dominant. Blue and Blue-Green ecosystems maintain similar proportions, and Brown ecosystems remain the least studied across both scales.

NbS in Europe range across a spectrum from protecting natural areas with minimal human intervention to modifying or creating entirely new green and blue infrastructure in urban environments. This flexibility allows NbS to function as both physical spaces – like parks or community gardens – and management practices – like rewilding or wetland restoration – offering multifunctional benefits such as cooling, stormwater control, and habitat creation. Yet, as observed in practice (see Appendix B), the current trend in Europe particularly emphasizes what is referred to as Type III interventions, i.e., the creation of new elements or redesigned spaces, that align with the compact city model. These include interventions like green roofs for urban heat mitigation or rain gardens for stormwater management, which are particularly effective for addressing localized needs for biodiversity support, climate adaptation, and social cohesion. Type I NbS that are conservation and restoration-based – such as those preserving remnant forests or rehabilitating degraded ecosystems – remain less common in urban environments, limiting opportunities to enhance landscape connectivity and protect existing biodiversity hotspots. As also observed in practice, a few projects integrate Type II restoration approaches.

This utilitarian model underscores the need for more balanced approaches that also prioritize long-term ecological integrity. Preserving

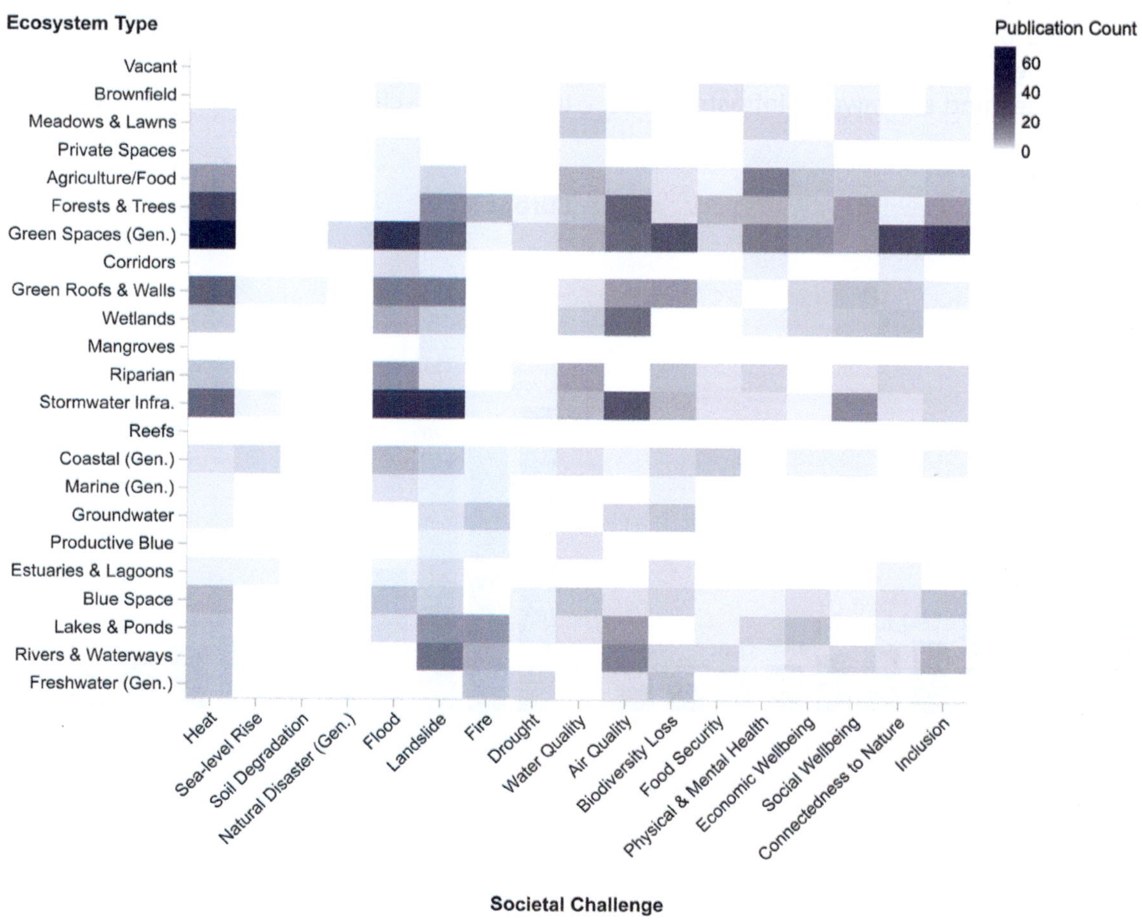

Figure 8: Co-occurrence of Ecosystem Types and Societal Challenges in European Academic Literature on Urban Nature-based Solutions.

natural areas may hinder urban expansion or redevelopment, creating a potential conflict with goals aimed at achieving biophysical, biodiversity-supportive cities. Additionally, the scale of NbS projects plays a crucial role in their effectiveness. Smaller projects, while more feasible, may lead to a problem of scale misfit, where solutions, though relevant in principle, fail to address the magnitude of urban challenges. This mismatch can result in interventions that are insufficient to mitigate issues such as climate change impacts, biodiversity loss, or urban flooding. Therefore, it is essential to carefully consider both the type and scale of NbS in urban planning to ensure they effectively contribute to sustainable and resilient city development.

Societal Challenges Addressed

The NbS interventions in the European region address diverse urban challenges (Figure 8), depicting the potential of NbS to deliver a wide range of benefits and co-benefits. The most studied ecosystem types are general green spaces and stormwater infrastructure, in relation to heat, flood and stormwater management. Green spaces are also studied in relation to water and air quality, health and wellbeing, which reflects how urban NbS provide opportunities for social revitalization in neighborhoods experiencing economic and social decline (Zwierzchowska et al., 2019), as well as access to spaces supporting mental and physical health in densely populated areas (Vujcic et al., 2017; Kolokotsa et al., 2020). Stormwater infrastructure also appears prominently, especially in relation to flood and water security challenges. Lastly, urban forests are also a predominant NbS type addressing primarily urban heat and air quality challenges.

The creation of greenery and regeneration of urban areas indirectly contributes to addressing other sub-challenges, including considerations of air quality, water management, social well-being, and socio-economic development. Climate change mitigation and adaptation are among the most important considerations when creating climate-resilient cities. Here, NbS such as green roofs, urban forests, and permeable surfaces, provide practical ways to enhance climate resilience by cooling

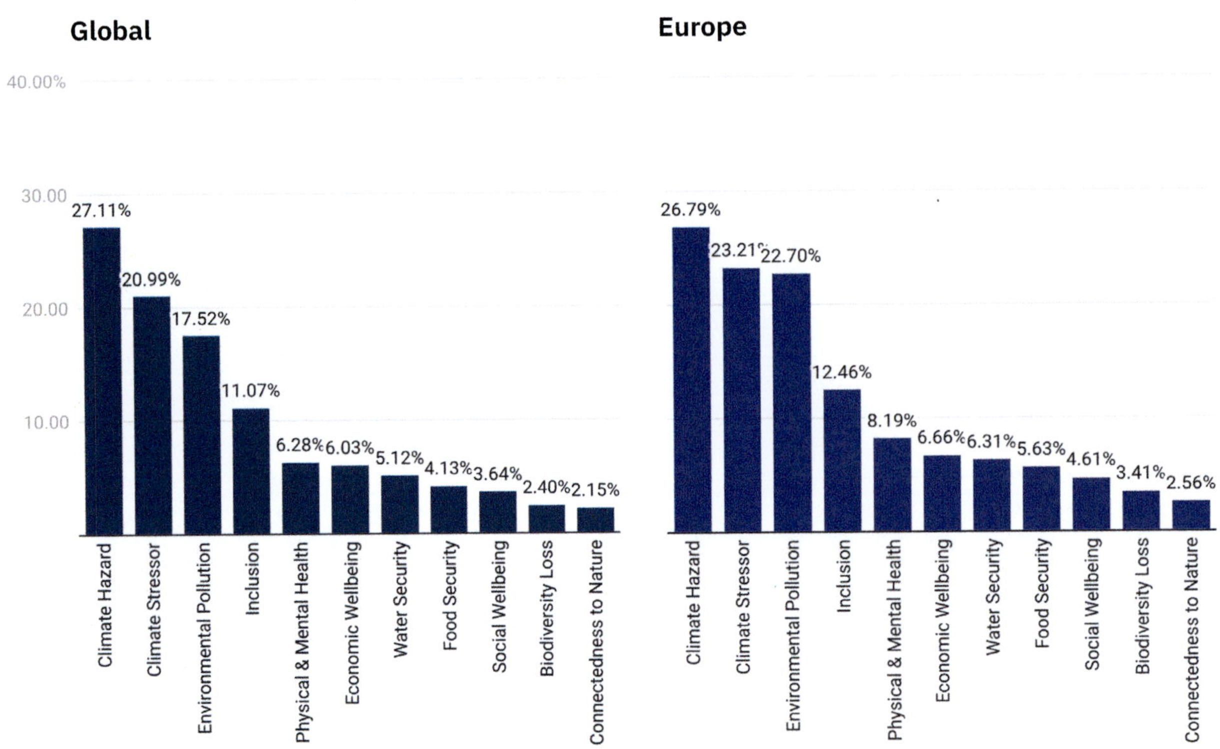

Figure 9: Proportion of Societal Challenges Addressed in Empirical Urban Nature-based Solutions Studies: Global vs. Europe.

urban areas, capturing carbon, and managing stormwater. The importance of (storm)water management in NbS is well-recognized, as it features prominently in high-level policies like UN-HABITAT and has influenced local strategies, such as the Copenhagen Climate Adaptation Plan. Similarly, sustainable water management and coastal protection are vital. Rising flood risks and sea levels can be addressed with NbS like rain gardens, wetlands, and coastal buffers, which naturally retain water and prevent erosion, safeguarding vulnerable urban and coastal areas.

When comparing the societal challenges addressed in empirical urban NbS studies globally and in Europe, several patterns emerge (Figure 9). The European urban NbS literature largely mirrors global trends, with climate-related challenges – particularly acute hazards like flooding and droughts, as well as long-term stressors such as sea level rise – being the most frequently addressed. This alignment reflects the central role of climate adaptation in NbS research worldwide and highlights Europe's significant contribution to shaping this agenda. Another commonality across both scales is the relatively limited attention given to biodiversity loss and human–nature connectedness, indicating persistent underrepresentation of these dimensions. In contrast, European studies place comparatively greater emphasis on challenges such as environmental pollution, public health, and social inclusion, likely reflecting the region's policy priorities and broader socio-political context.

While urban NbS can play a significant role in addressing a wide range of societal challenges, it is crucial to recognize that their effectiveness varies depending on the nature of the challenge. Some challenges have a more direct and well-established link to NbS, while others require further research to fully understand their interactions and dependencies. For example, NbS can directly contribute to climate adaptation, where their benefits are well-documented. However, their role in addressing systemic social and economic issues – such as poverty, socio-economic vulnerability, and segregation – requires more nuanced integration with broader infrastructural, economic, and political structures. While environmental justice and anti-green gentrification considerations are essential in NbS planning, NbS alone cannot overcome the deep-rooted political-economic structures that drive inequality (Lampinen et al., 2021; Kato-Huerta and Geneletti, 2022; Anguelovski and Corbera, 2023). Similarly, while NbS contribute to public health and well-being by improving air quality, providing access to green spaces, and mitigating urban heat, their impact on broader social determinants of health – such as income, housing, and education – remains less direct and requires further study.

3.1.3 | Actors, Partnerships, and Finance

KEY FINDING 3:

NbS practices are increasingly managed through multi-actor partnerships while being supported by new funding models, largely catalyzed by Horizon 2020 and Horizon Europe and complemented by independent initiatives from various government levels.

NbS projects in Europe demonstrate a complex set of partnerships fundamentally shaped by EU funding instruments, which explicitly require multi-stakeholder engagement. For this reason, urban NbS practices in the region involve diverse categories of partners, including governmental bodies (particularly prominent at municipal levels), academic institutions, SMEs/businesses, and civil society organizations. The high prevalence of governmental partners, especially local authorities and municipalities, reflects the policy-driven nature of European NbS initiatives, with these authorities playing crucial roles in regulation, urban planning, policy support, and actor mobilization. In some case studies, public partner collaboration is enriched by a diverse network of institutionalized NGOs such as WWF and other local NGOs.

EU funding mechanisms, particularly through Horizon Europe, play a crucial role in fostering collaborations by bringing together diverse academic, governmental, and private stakeholders (see Box 1). Many projects are funded through Horizon Europe because it provides explicit funding specifically designated for NbS-related research, implementation, and scaling. NetworkNature is a key initiative in this space. As an EU-funded Coordination and Support Action, it connects over 100 Horizon 2020, Horizon Europe, and EU Mission projects on NbS, representing a significant share of EU investment in NbS research and innovation. It provides a central coordination platform for the European NbS R&I landscape and supports the EU Research and Innovation Roadmap through its Task Forces and thematic work on policy, governance, finance, and standardization. However, other EU funding programs (see for example LIFE, INTERREG) also support

The European urban NbS literature largely mirrors global trends, with climate related challenges – particularly acute hazards like flooding and droughts, as well as long-term stressors such as sea level rise – being the most frequently addressed. This alignment reflects the central role of climate adaptation in NbS research worldwide and highlights Europe's significant contribution to shaping this agenda.

initiatives that contribute to NbS, even if they are not exclusively framed around NbS. These programs often fund projects related to biodiversity conservation, climate adaptation, and sustainable land and water management, complementing the more targeted NbS funding streams under Horizon Europe.

Specific NbS interventions have been implemented across Europe through public funding in programs such as Horizon 2020 and Horizon Europe (European Research Executive Agency, 2025). The predominance of EU-origin cases can be attributed to EC funding schemes providing access to abundant, detailed information and addressing various environmental and social challenges. For example, the EC invested 887 million euros (about 1.050 billion USD) in NbS-related initiatives, thus supporting 105 projects (in both urban and non-urban areas). Moreover, multiple financial mechanisms have emerged, often in collaboration with these programs, including joint partnership initiatives involving EU Member States and

BOX 1 **Actors and Partnerships in Sofia (Bulgaria)**

The six-year URBiNAT project (2018-2024) focused on regenerating and urbanizing deprived social housing by implementing NbS. The case study in Sofia presented a complex interplay between traditional municipal administration and innovative participatory approaches. Sofia Municipality led the initiative at its core, with involvement from both central and district-level administrations, highlighting a multi-tiered local government engagement. This collaboration was complemented by a partnership with the University of Architecture, Civil Engineering and Geodesy, bringing academic expertise into urban planning. The project's governance model emphasized stakeholder involvement and citizen engagement through the Living Lab approach in Nadezhda district, aiming to co-create the Healthy Corridor concept. The participatory approach sometimes encountered friction with established structures, balancing the need for administrative procedures with flexibility. The municipal team demonstrated adaptability as the project progressed, evolving to meet changing needs. While communication challenges remained within the team and with external partners, this governance model aims to bridge traditional frameworks with innovative practices by navigating the complexities of inter-sectoral coordination and integrating diverse actors in planning and implementing NbS.

seed funding from private foundations and companies. According to the NATURVATION Atlas, nearly 75 percent of NbS projects are financed through public sources, such as direct funding, subsidies, and public budgets.

The cross-sectoral approach to implementing NbS – emphasizing collaboration among stakeholders through financial, regulatory, and other instruments (European Commission, 2023b) – has helped catalyze private sector investment in NbS. The European Investment Bank's 2023 report, Investing in Nature-based Solutions (Hudson, Hart and Verbeek, 2023), explains that while public funding from the EU, national, and local levels is still the main source, interest from private investors – especially banks and financial institutions – is steadily growing. Despite private finance making up the majority of funding in only 3 percent of NbS initiatives, it can play a critical role in unlocking blended finance models. These multi-sectoral mechanisms combine public and private resources, aligning financial returns with ecological benefits (see Box 2, appendix EU_19).

BOX 2 **Seed private funding in Rumon (Italy)**

Volunteers at a public garden in Rumon. Antonio Luizo and Elisa Debbi, 2025.

The case study in Finale Emilia, Italy, is an interesting example of private and mixed financing built on a local active citizenship effort (see appendix EU_19). Rumon began as a self-funded project, supported by 4,000 EUR (4,212 USD) in private donations and an 800 EUR (842 USD) contribution from Manitese social cooperative. A local bank foundation, Fondazione Cassa di Risparmio of Mirandola, contributed an additional 4,000 EUR (4,212 USD) , enabling the purchase of an irrigation system and event support. The project also received private donations. The 'Free Trees' festivals (November 2022 and April 2023) funded two rounds of tree plantings through the regional 'Roots for the Future' program, which offers free young trees to Emilia Romagna residents. In 2024, Rumon joined Manitese cooperative, gaining access to official funding opportunities as a member of the National Third Sector Register (RUNTS), while still relying on volunteer support.

3.1.4 | Inclusion and Participation

KEY FINDING 4:

Participatory approaches to NbS are increasingly common across Europe, but often fall short of meaningful co-creation, with limited engagement of marginalized groups.

In Europe, mechanisms for inclusion are widely recognized as a critical component of NbS. Co-creation helps ensure that different voices are heard and that solutions match local needs (Frantzeskaki and Kabisch, 2016; Haase, 2017). Without a critical reflection on the diversity of voices and their degree of influence, NbS can miss the mark or even worsen inequalities, especially if engagement is only symbolic and does not influence decisions (Dushkova and Haase, 2023). Setting up meaningful participation, however, takes time, resources, and trained staff. It also requires funding schemes that support long-term involvement. While social and cultural benefits of participation are often mentioned, the difficulties – such as tensions in group dynamics – are still not well studied (see Section 3.6). Even so, advanced practices are emerging, showing how inclusive processes can make NbS more just and transformative.

Participatory approaches aim to incorporate diverse stakeholder perspectives into planning and implementation processes, ensuring that NbS address local needs and priorities (Raymond, Frantzeskaki, et al., 2017). Common mechanisms include public consultations, stakeholder advisory groups, participatory workshops, and deliberative forums. However, without genuine co-creation and participation, there is a risk that NbS fail to adequately reflect local needs, leading to reduced effectiveness, public resistance, or unintended inequities in outcomes. Tokenistic participation, where involvement is limited to superficial feedback collection without meaningful influence on decision-making, undermines trust and the legitimacy of NbS initiatives. Moreover, organizing systematic and meaningful participation is inherently complex and time-intensive, requiring tailored funding schemes and capacity-building efforts for facilitators and administrators. Social justice, cohesion, and equity are frequently considered essential to well-planned NbS, and cultural and social aspects are often highlighted as co-benefits, particularly through co-creation activities (Castellazzi, Hakkarainen and Raymond, 2024). Still, they are not analyzed in depth, and challenges or tensions are usually overlooked (as explored in section 3.6). However, more advanced practices are emerging where stakeholders, including vulnerable groups, actively shape NbS strategies.

The principles of social inclusion, justice and participation are the cornerstones of procedural justice and innovative co-creation processes. Knowledge co-creation, as articulated through phases of co-diagnostics, co-design, co-implementation, co-monitoring/evaluation, and co-amplification/replication, provides a structured framework for engaging diverse stakeholders and addressing social disparities (Clever Cities, 2021; European Commission et al., 2023). By involving vulnerable groups, as seen in Belfast's community-led garden initiatives (see Box 3), Zagreb's therapeutic garden or Glasgow's long-term environmental planning, co-creation enables historically marginalized communities to shape urban interventions actively, aligning NbS with their specific needs and priorities (Baud et al., 2021; European Commission, 2023a).

BOX 3 Co-creation in Belfast (Ireland)

People coming together at a community garden in Belfast. Emma Campbell, 2025.

Participants ideating local NbS structures. Emma Campbell, 2025.

Different kinds of people led different stages of co-creation in the UPSURGE project's Belfast demonstrator. At the beginning of this project, Queen's University, Belfast's Architecture department devised engagement approaches for the co-diagnostics and co-design stage. Once the site was co-designed, Belfast City Council developed closer working relationships with key collaborators, such as the community group Friends of the Field. These relationships focused on co-designing individual plots, such as for the planned community garden, and exploring opportunities for collaborative construction (co-implementation), citizen-science (co-monitoring) and sharing responsibility for site management (co-governance). Over time, Friends of the Field were supported in leading bottom-up co-creation engagement through on-site events. For example, in March 2024, they hosted an open day for a community garden with activities such as gate painting, willow-weaving for raised beds, and bamboo hut construction.

Designing and implementing NbS in a post-conflict city like Belfast has proven complex. The emerging concept of co-creation has required significant trial and error, as the team navigated the need for diverse skills and roles throughout the project's phases. While shared responsibility in facilitating co-creation was intended to foster inclusivity, it sometimes led to frustration and confusion among collaborators, who questioned the project's leadership and coordination. Despite these challenges, one effective approach has been to empower bottom-up actions directly on-site. This grassroots involvement has granted citizens a tangible sense of agency, allowing them to engage meaningfully with the project. Hands-on participation has positively influenced governing bodies too by encouraging them to accelerate change, decant power structures and open pathways for future NbS initiatives.

Co-creation approaches emphasize working together to develop solutions, drawing on the idea of collaborative innovation promoted by the EU. This approach is held to ensure a fair and inclusive transition toward climate resilience (European Commission, 2013). As projects like UNaLab and Connecting Nature illustrate, co-creation goes beyond consultation, embedding citizens directly in decision-making, thus decentralizing power and enhancing social inclusion (Hölscher et al., 2023; *UNaLAB Toolkit*, 2024). For example, initiatives like Intercommunal Dune Management in Gâvres utilize citizen monitoring to increase transparency and trust. This shows that co-creation fosters awareness and mitigates tensions around resource allocation or environmental interventions (Wilk et al., 2020; Hölscher et al., 2024).

Procedural justice in NbS is further promoted by expanding models of engagement to include stakeholders such as scientists and researchers who advocate for environmental perspectives often overlooked in urban planning (Vignola et al., 2009; Carayannis, Barth and Campbell, 2012; Demuzere et al., 2014). This diversification ensures that NbS projects are ecologically informed, sustaining human and environmental health. In light of reduced public budgets for the operational management of green spaces, some projects have explored the potential for non-governmental actors to take on roles in place-keeping. The Krakow Green Krupnicza case exemplifies this and demonstrates how engaged local citizens can serve as stewards and maintainers of these projects over time. However, for co-creation to better represent the needs of diverse groups, challenges like funding, long-term governance, and public acceptance must be addressed (Schmalzbauer, 2018; International Union for Conservation of Nature, 2020; Seddon et al., 2020; European Commission, 2023a).

The practice of urban NbS in the region relies on various participatory mechanisms that integrate non-governmental actors and civil society into NbS decision-making processes. One example is the application of Module III, "Deliberative Co-Creation", from the INTERLACE assessment framework (Langemeyer et al., 2022). This approach was used to elicit social demands of different social groups for ecosystem services, informing the restoration of the Congost River in the city of Granollers as part of the CO-CO-NAT 2025 project. Another example is the use of a Public Participatory Geographic Information Systems (PPGIS) surveys. This method engaged local residents to identify social values assigned to specific places and potential justice concerns related to implementing urban NbS, as demonstrated in the Hørgården case (Maurer et al., 2023).

Despite the attention to co-creation processes, the question of 'NbS for whom?' remains an active area of attention. While involving people in nature-led projects is often easier than other urban projects, a broad spectrum of people must be involved (Clever Cities, 2021). Research shows multiple disparities and inequities concerning NbS planning, reflected in an underrepresentation of people that may require additional efforts to engage with, such as women and ethnic minority groups (Kato-Huerta and Geneletti, 2023), as well as youth and the elderly. Those instigating co-creation processes must assess who might be missing from the room and tailor approaches to expand engagement.

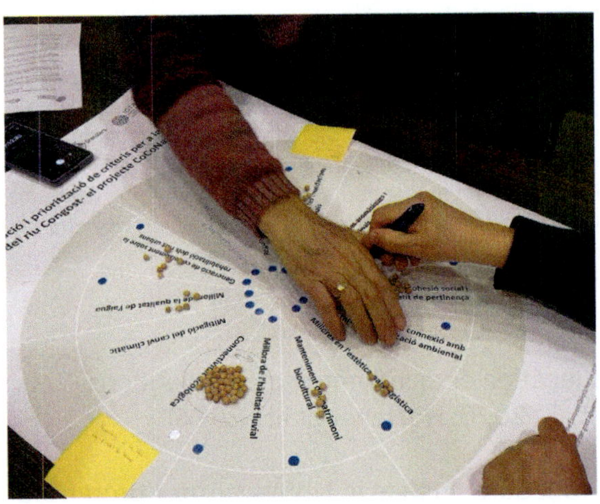

Participatory workshop with stakeholders to identify the evaluation criteria for the CO-CO-NAT 2025 project in the city of Granollers, Spain. Oriol Garcia-Antúnez, 2025.

Horizon 2020 invested 665 million euros (818 million USD) in NbS-related initiatives, thus supporting 76 projects mainly focused on urban areas. Moreover, multiple financial mechanisms have emerged, often in collaboration with these programs, including joint partnership initiatives involving EU Member States and seed funding from private foundations and companies.

Intergenerational participatory workshop in Granollers, where youth and elderly participants expressed their priorities, aspirations, and expectations for the CoCoNat 2025 project using a variety of participatory tools. The City of Granollers, 2025.

3.1.5 | Policies, Legal Frameworks, and Urban Strategies

KEY FINDING 5:

Europe has developed a dense web of policy and regulatory frameworks to mainstream NbS, but fragmented governance and insufficient capacity for enforcement at the local level hinders effective implementation.

Governance around climate change and NbS in Europe is often described as “multi-level” due to its involvement in local, regional, and national decision-making, with public and private actors participating. The EU’s multi-level governance approach to climate change and NbS relies on coordinating policies and planning tools across EU, national, regional, city, and local levels. Each level contributes uniquely to implementing NbS, with targeted strategies that adapt broader EU goals to local contexts.

The EU commitment to NbS is strongly reflected in the European Green Deal, a set of strategic initiatives to promote sustainability and resilience in Europe. A central goal of the Green Deal is to achieve climate neutrality across the EU by 2050 and integrate NbS into critical sectors, including agriculture, forestry, and urban development. Moreover, the Green Deal’s emphasis on NbS is underscored by diverse targets for ecosystem restoration, biodiversity protection, and climate resilience. Within the EU, treaties establish legally binding obligations and rights for all member states to ensure the effective functioning of the union and guide secondary legislation, such as directives and regulations. Regarding environmental matters, legally binding frameworks include the Water Framework Directive, the Habitats Directive, and the new Regulation on Nature Restoration. On the other hand, the non-legally binding policy framework of the European Green Deal set strategic goals and priorities for sustainable development and climate action. While these frameworks do not impose legal obligations, they strongly influence member states’ policies and encourage alignment toward shared

More than two hundred companies, associations, and non-governmental organizations join to urge national leaders to reaffirm their dedication to the European Green Deal. European Council/European Union, 2024.

environmental targets, such as achieving carbon neutrality by 2050.

Key policies advancing NbS implementation in Europe reflect the commitments of the Green Deal, including the Climate Law, Climate Adaptation Strategy, 2030 Biodiversity Strategy, Forest Strategy, Farm to Fork Strategy, Circular Economy Action Plan, Zero Pollution Action Plan, and EU Sustainable Finance Strategy. Among these, the newly introduced Nature Restoration Regulation has become a cornerstone of the EU's environmental agenda, solidifying NbS as central to European climate and biodiversity goals.

However, while these legislative and strategic frameworks provide a strong foundation, significant gaps in uptake and implementation persist. These gaps hinder the achievement of intended goals, as many NbS remain underutilized or improperly executed. Without adequate funding mechanisms, capacity building, and administrative support for participatory processes, these ambitious policies risk falling short of their potential, limiting their impact on climate resilience and biodiversity restoration (Hering et al., 2023). Addressing these barriers is vital to bridging the gap between policy and practice, enabling NbS to fully contribute to Europe's environmental objectives.

At the highest policy level, the EU Urban Agenda provides a framework for sustainable urban development, helping cities integrate NbS into urban planning and as infrastructure. It encourages member states to align with the EU's sustainability and climate neutrality goals, setting a guiding structure for local adaptations. Several EU countries have enacted national policies that support the implementation of NbS through the Urban Agenda. In Spain, the Spanish Urban Agenda provides a foundation for sustainable urban planning in municipalities like A Coruña. Similarly, Denmark's national laws inform NbS initiatives such as the Hørgården case (see Box 4), focusing on ecological resilience and urban green spaces. In France, the GEMAPI Law establishes a framework for managing aquatic environments and flood prevention, aiding coastal areas like Gâvres in using NbS to reduce flood risks and enhance biodiversity.

Authorities also tailor regional policies to meet local needs while supporting national and EU goals. For example, the Ruhr Regional Association coordinates green infrastructure projects in Gelsenkirchen, fostering cross-municipal cooperation for ecosystem restoration. In Italy, the Emilia Romagna Regional Council leads initiatives that promote NbS for regional resilience and disaster preparedness. Cities are embedding NbS into urban planning through comprehensive strategies like Lisbon's Green Plan, Brussels'

Climate Action Plan, and Glasgow's Open Space Strategy, which enhance biodiversity, green spaces, and climate resilience. At a local level, Sofia's development plan integrates NbS in neighborhoods to improve community well-being and environmental health. Complementing these efforts, green point systems are emerging as tools to measure and incentivize sustainable urban development. These systems often align with NbS principles and offer an opportunity to standardize and expand their application, bridging strategic planning with actionable outcomes. Together, they illustrate a cohesive approach to integrating NbS at both strategic and practical levels.

While these legislative and strategic frameworks provide a strong foundation, significant gaps in uptake and implementation persist. Without adequate funding mechanisms, capacity building, and administrative support for participatory processes, these ambitious policies risk falling short of their potential.

BOX 4 **Multi-level Policies and Financial Mechanisms for NbS (Copenhagen)**

Flowers blooming in a Hørgården garden. Lft3b via Wikimedia Commons, 2011.

View of the housing area from the garden. Selveje Netværket, n.d.

The case study in Hørgården, Copenhagen, is an excellent example of multi-level governance of NbS for addressing social-ecological challenges at the neighborhood scale. Local municipal and national efforts have been coordinated for over 10 years to ensure spatial and social planning coherence to (1) open and densify the neighborhood, (2) to support urban greening and biodiversity initiatives, and (3) to finance stormwater management in a socially and economically marginalized neighborhood. Hørgården is a centrally-located not-for-profit housing area that provides affordable housing to approximately 1500 lower-income residents in Copenhagen. It has been identified as a marginalized area by the municipality and the national government. In 2018, this designation unlocked funding (9 million EUR; 10.6 million USD) for an infrastructure plan to support the engagement of residents in "opening up" the neighborhood spatially to reduce the social segregation of the area through new green corridors and connections to the surrounding area. NbS serve both to enhance the biodiversity and aesthetics of Hørgården's green commons and as a vehicle for engaging local residents and fostering a sense of stewardship and social cohesion. The implementation process of the Hørgården Infrastructure Plan is being coordinated with a stormwater management project driven by the utilities company, HOFOR. In line with the Copenhagen Climate Adaptation Plan (2012), Hørgården has been slated for a major stormwater adaptation process to both absorb on-surface storm water into the green commons and replace the current underground sewer system with new pipes to separate sewage from rainwater runoff (Maurer et al., 2023). Therefore, the governance of climate adaptation in Hørgården exemplifies how national and municipal planning tools can collectively promote implementing NbS.

3.2 Challenges to Implementation

As demonstrated by the included case studies, European urban areas have accumulated a wealth of local experiences with NbS, primarily through pilot projects where these solutions are tested at relatively small spatial scales, such as individual sites or neighborhood levels. However, scaling out remains a key challenge. While NbS have demonstrated their potential in localized settings, their broader deployment as an integral urban planning strategy at the city and metropolitan scales is still an ongoing challenge. This lack of widespread adoption stems from multiple interconnected barriers hindering uptake and diffusion (Sarabi et al., 2020). These obstacles arise from the complex interplay of urban environments' social, ecological and technological characteristics (McPhearson et al., 2016). Identifying and understanding these barriers is crucial for the effective implementation of NbS. To better understand why NbS are not yet widely used at larger scales, we need to look not just at technical issues but also at these deeper layers. Based on the case studies, we identify four main types of barriers: gaps in knowledge and technical capacity, financial and economic constraints, mismatches with public expectations, and problems with governance and coordination.

3.2.1 | Technical and Knowledge Barriers

KEY FINDING 6:

Technical gaps in design, integration, and long-term monitoring undermine the resilience of NbS. Without standardized guidelines, sufficient capacity, or data on how benefits evolve over time, many projects in the region struggle to sustain impact and validate effectiveness. These challenges are compounded by conceptual ambiguity – without a clear and operational definition, practitioners sometimes face difficulties in translating NbS into coherent design standards, performance benchmarks, and maintenance protocols.

While there has been significant progress in defining and exploring the potential of NbS, practitioners face challenges in translating this concept into actionable strategies. Experience from urban NbS practice across the region points to persistent technical barriers – ranging from ambiguous definitions to difficulties integrating NbS with existing infrastructure, and insufficient monitoring of their co-benefits. Without a clear, universally accepted definition, efforts to plan and implement NbS projects can run the risk of becoming fragmented. Stakeholders may work from differing interpretations, leading to complications in communication and coordination both within teams and with the public. This misalignment not only hinders the ability to secure support but also jeopardizes the long-term success of projects. Addressing this ambiguity is critical to bridging the gap between conceptual understanding and effective implementation.

Another conceptual challenge relates to the way in which NbS have been operationalized and institutionalized in Europe – particularly through the European Commission's research and innovation funding programs such as Horizon – has shaped how the concept is understood both within and beyond Europe. The Commission's agenda, emphasizing priorities like cost-effectiveness and infrastructure planning, has influenced the framing of NbS, notably through the integration of economic development objectives. This marks a departure from earlier definitions, such as that of the IUCN, which emphasized ecological and social dimensions. While this European approach may work well for local challenges, it might not fit other social and environmental contexts. This shows the importance of thinking critically about how NbS are defined and the political ecology driving this conceptualization.

The Commission's agenda, emphasizing priorities like cost-effectiveness and infrastructure planning, has influenced the framing of NbS, notably through the integration of economic development objectives. This marks a departure from earlier definitions, such as that of the IUCN, which emphasized ecological and social dimensions.

Compounding these challenges are the barriers associated with implementing NbS in urban areas, where historical legacies and systemic lock-ins create significant constraints. Unlike blank canvases, cities are often characterized by highly engineered environments that limit the feasibility of restoring natural processes and adapting to or integrating with pre-existing structures. For instance, NbS aimed at stormwater management often contend with intricate underground infrastructure that complicate both technical implementation and economic viability. Copenhagen's Grønne Veje project exemplifies this issue as the use of rain beds for managing stormwater was impeded by the dense network of underground electricity, sewage, and internet systems. Similarly, Krakow's Green Krupnicza project highlighted the difficulty of integrating permeable streets into a pre-existing web of roads and drainage infrastructure. In both cases, these complexities required specialized expertise throughout the process – from planning to construction and ongoing maintenance.

The Grønne Veje project in Copenhagen. Oriol Garcia-Antúnez, 2025.

Unlike grey infrastructure, NbS lacks uniform standards across many cities. Without clear technical guidelines, ensuring consistency, safety, and efficiency in NbS projects becomes a significant challenge. Each city may be forced to develop its own protocols, leading to inefficiencies, higher costs, and reduced confidence among stakeholders. For example, in Copenhagen, the Grønne Veje project case study faced challenges due to the lack of standardized guidelines for implementing Sustainable Urban Drainage Systems (SUDS) as part of its NbS strategy. This lack of standardization also reflects the overall technical unreadiness of administrative staff and professionals working with NbS, which are required to integrate siloed themes across social, ecological and technological realms into single projects.

Another relevant challenge is the monitoring of their outcomes. Long-term monitoring is critical for urban NbS, as their impacts often take time to fully materialize. However, this need for extended observation contrasts with the shorter time frames typically favored by political decision-making, which prioritizes quick and tangible results (Sarabi et al., 2020; Linnerooth-Bayer et al., 2023). Additionally, the measurement of NbS benefits must be conducted locally, as social factors (e.g., building location and height) and ecological conditions (e.g., climate variability) can significantly influence the uptake and performance of these solutions (Grilo et al., 2020). Despite these necessities, much of the knowledge on NbS functionality and benefits remains scattered (Nesshöver et al., 2017). In fact, the majority of projects reviewed in this report fell short in the area of monitoring, highlighting a broader trend of neglecting ongoing monitoring and maintenance compared to the emphasis placed on implementation and other initial project phases. For example, the UrbanGreenUp project in Liverpool, which implemented green infrastructure for climate mitigation and biodiversity enhancement, encountered challenges due to the absence of established guidelines for quantifying and monitoring ecological benefits over time. This lack of monitoring is crucial to address, as it plays a vital role in validating the effectiveness of NbS and safeguarding them against risks of greenwashing and the overselling of benefits (Li et al., 2025).

3.2.2 | Economic and Financial Constraints

KEY FINDING 7:

Despite growing evidence of their long-term cost-effectiveness and resilience benefits, NbS are still perceived as high-risk investments, which limits private sector engagement.

NbS are often perceived as high-risk investments due to uncertainties surrounding their long-term financial returns. Unlike traditional grey infrastructure, such as concrete drainage systems, the economic benefits of NbS – such as ecosystem services, climate resilience, and improved public health – are not always immediately visible or easily quantifiable. This perception can discourage stakeholders from committing resourcves. For instance, Lisbon's Urban Agriculture Strategy aims to establish allotment gardens that promote community resilience and food security. However, maintaining these gardens, particularly to ensure accessibility for low-income communities, requires sustained investment. Uncertainty about long-term public interest and the gardens' resilience over time complicates efforts to secure funding.

This challenge reflects a broader issue: the need for more research into the relative costs and benefits of NbS in different settings. Equally important is understanding who bears the costs and who reaps the benefits, as this often varies by context. For example, while some stakeholders may benefit from improved ecosystem services or increased property values, others may bear the financial burden of maintenance or implementation. These dynamics underscore the importance of contextualized, place-specific research to assess how NbS can deliver equitable and effective solutions (European Commission, 2020).

A struggling garden in Lisbon. Filipa Grilo, 2025.

3.2.3 Public Perceptions of and Aspirations for Urban Nature-based Solutions

KEY FINDING 8:

Overcoming skepticism – rooted in concerns over effectiveness, maintenance, or spatial trade-offs – depends on how well interventions align with local priorities, cultural values, and everyday experiences.

Public perceptions of urban NbS influence their social acceptability – reflected in public approval and positive attitudes – and their political acceptability, which refers to the political willingness and commitment to support and invest in these initiatives (Sari et al., 2023). Politically, stakeholders may hesitate to adopt NbS practices due to uncertainties and concerns about potential financial losses (Kabisch et al., 2016;

Linnerooth-Bayer et al., 2023). This hesitation often stems from the misconception that NbS, particularly their maintenance, entails higher costs than conventional grey infrastructure (Lorenzoni, Nicholson-Cole and Whitmarsh, 2007). Additionally, conflicting interests among decision-makers and siloed governance approaches can exacerbate these challenges (Clar, Prutsch and Steurer, 2013). However, driven by the EC's advocacy for NbS, growing experience and knowledge about their benefits, and increasing international momentum, these political misconceptions are gradually giving way to greater political acceptance, which is critical for the long-term success of NbS.

In terms of public attitudes, limited awareness and negative perceptions remain significant barriers to NbS adoption (Wamsler et al., 2020; Linnerooth-Bayer et al., 2023). For example, a study in Helsinki, Finland, revealed that perceptions of ineffectiveness – beliefs that urban NbS cannot adequately address key challenges – combined with low ecological literacy contributed to a lack of support for carbon-smart NbS aimed at enhancing carbon sequestration and storage (Garcia-Antúnez et al., 2023). This dynamic is also evident in the case studies examined in this report. In Eindhoven, for example, a workshop identified barriers to NbS adoption, with one of the most critical being the limited dissemination of information to citizens regarding the functionality and performance of NbS. Public risk perceptions also play a significant role in shaping the approval of urban NbS, particularly in the context of "soft" solutions for managing hydro-climatic risks and weather-related hazards such as river flooding, stormwater management, and coastal erosion. Misconceptions persist that "soft" NbS offer lower levels of protection against infrastructure damage, human safety risks, and economic loss compared to "hard" engineering approaches (Tunstall et al., 2000; Andersson et al., 2021). Although these misconceptions are gradually receding, the case of Granollers demonstrates how local concerns about flood risks and doubts about the effectiveness of NbS can still hinder their social acceptability at the community level.

A key factor influencing social acceptance of NbS is the alignment with local aspirations for urban greenery. As cities densify and per capita green space declines (Haaland and Van Den Bosch, 2015), residents may resist NbS strategies if they perceive them as failing to meet their expectations for accessible, multifunctional green spaces (Garcia Antúnez, 2024). When interventions do not reflect community values, they risk low public acceptance (e.g., Lampinen et al., 2021, 2023). At the same time, urban densification intensifies competition for space: the demand for housing and recreational areas often takes priority, constraining opportunities for large-scale NbS implementation. Overcoming these spatial challenges requires integrated, multifunctional approaches that maximize co-benefits within limited urban space. A strong example is the SPARK project in Aarhus, Denmark, which successfully integrates multiple policy objectives. Designed primarily for stormwater management, it also serves as a recreational space for residents and an outdoor rehabilitation area for a nearby center. This project illustrates how well-planned NbS can address multiple urban challenges while optimizing the use of scarce space.

Social acceptance is crucial for securing sustained political support for urban NbS, particularly when these initiatives become politically contentious. Public perceptions significantly impact the stability of NbS projects; high levels of acceptance can reduce polarization and establish them as valued community assets, whereas low support increases their vulnerability[3]. This feedback loop between social and political acceptance shows how these two are interconnected and equally essential for urban NbS interventions' long-term success.

[3] *For example, in Barcelona, the Superblocks initiative – a project to prioritize pedestrians and green spaces by restricting vehicle traffic in designated urban areas – has sparked intense public and political debate. While many support its environmental and social benefits, others criticize it for causing traffic disruptions and impacting local businesses, making it a highly politicized urban transformation.*

Unlike traditional grey infrastructure, such as concrete drainage systems, the economic benefits of NbS – such as ecosystem services, climate resilience, and improved public health – are not always immediately visible or easily quantifiable.

SPARK project implemented in Aarhus, Denmark. Kristine Jensen, 2022.

3.2.4 | Governance and Institutional Barriers

KEY FINDING 9:

Many cities struggle to operationalize NbS without coordinated governance frameworks, clear procurement pathways, and supportive legal standards.

The successful implementation of NbS is heavily reliant on effective governance frameworks. Governance challenges for NbS are often rooted in fragmented responsibilities, regulatory gaps, a lack of policy support, and limited coordination among stakeholders. Moreover, many cities lack clear procurement practices, regulations, and standards to mitigate financial liabilities associated with NbS (Sarabi et al., 2020). For example, in Zagreb's proGIreg project, procedural challenges led to delays and cancellations of original implementation plans, undermining confidence in NbS as a viable solution.

In many urban settings, the governance of NbS is divided across multiple departments or agencies, each with its own priorities, mandates, and operating procedures. This fragmented responsibility can create silos, where different teams work in isolation without a cohesive strategy. Siloed decision-making is particularly problematic for NbS, as these solutions often intersect with various domains – such as water management, public health, urban planning, and environmental protection – that require integrated approaches. For example, in Lisbon's Urban Agriculture Strategy, multiple city departments establish and maintain allotment gardens meant to provide accessible green spaces to low-income communities. However, fragmented responsibilities across departments can lead to delays in decision-making, with departments sometimes prioritizing their agendas over the collective goals of the NbS project. Another clear example of administrative siloing can be seen in the "Green Street" project in Copenhagen, Denmark. In this case, legal and financial compartmentalization within the water utility company hindered meaningful engagement with the local community. Economic resources were solely directed toward technical stormwater management infrastructure, limiting opportunities for broader collaboration and community co-production. These examples demonstrate how administrative silos can obstruct efforts to generate co-benefits, particularly in projects that require strategic and integrated governance approaches.

Regulatory and policy challenges further complicate NbS governance, as outdated or insufficient regulations create uncertainty for implementers and deter investment. For instance, Liverpool's UrbanGreenUp project, which seeks to create green spaces for climate mitigation, has struggled due to a lack of supportive policies and legal guidelines for climate-focused NbS. Without formal policy frameworks that recognize the role of NbS in climate adaptation, such initiatives face difficulties in securing long-term funding and institutional backing.

Finally, effective NbS implementation requires the involvement of a wide range of stakeholders, including government agencies, private sector partners, local communities, and non-governmental organizations. However, coordinating these actors presents significant challenges, particularly when their levels of influence, expertise, and commitment vary. In the UNaLab project in Eindhoven, the Netherlands, stakeholder involvement was critical, but aligning their diverse interests and ensuring long-term commitment proved difficult. Limited coordination led to delays and compromises that impacted the quality and effectiveness of the NbS.

3.3 Bright Spots

Here, we highlight NbS bright spots in European cities, which are unique projects, approaches, and partnerships that go beyond standard practices and illustrate pathways for scaling up urban NbS. These examples showcase how innovative monitoring, financing, participatory processes, and governance can help embed NbS as integral elements of urban transformation.

Technical & Knowledge Bright Spots

Monitoring the long-term success and performance of NbS projects remains limited, hindering the synthesis of knowledge and experience that is key for the outscaling and upscaling of NbS. The CLEARING HOUSE Horizon 2020 project in Germany serves as an inspirational bright spot, demonstrating how long-term ecological data collection can help bridge this gap. By establishing two Marteloscopes – fully inventoried forest plots – at the Rheinelbe industrial forest and the former Zeche Hugo coal mine, the project tracks various ecological attributes, including tree species, diameter, above-ground biomass, and ecological value (e.g., presence of microhabitats).

The Rheinelbe Industrial Forest in Germany. Astrid Hamm, 2006.

This data can be used to assess the ecosystem services provided by this NbS over a long time series, as the post-industrial site undergoes natural succession toward a more mature forest state. Implementing such long-term monitoring schemes – especially if initiated before NbS interventions – can be highly valuable for assessing the ecological impacts of NbS projects and thus guide improved NbS planning (particularly for NbS types I and II). Moreover, these monitoring sites can serve as platforms for educating and engaging local stakeholders in understanding the multifunctional benefits of urban forests and fostering collaborative decision-making. This approach not only enhances technical understanding but also builds public trust, facilitating more effective implementation of NbS in urban environments.

Economic & Financial Bright Spots

While NbS often involve specific costs – such as vegetation maintenance and ecosystem monitoring – they can deliver significant long-term benefits, including enhanced resilience, improved aesthetics, and reduced flood risk (Russo and Cirella, 2021). A key innovation of the Grønne Veje project in Copenhagen lies in how nature-based rain beds were formally integrated into the city's grey infrastructure system. Through the Cloudburst Management Plan, the City of Copenhagen allocated 900 million DKK (144 million USD) to HOFOR, the public water utility, to implement these green streets as technical drainage infrastructure. This legal and institutional framing allowed rain beds to be funded through public utility budgets – typically reserved for conventional grey infrastructure – thereby ensuring their construction and long-term maintenance with taxpayer money. As a result, rain beds could be deployed even on privately owned streets, relieving landowner associations of financial and technical responsibilities. This integration into the grey infrastructure domain not only secured the long-term sustainability of the NbS but also marked a significant step toward normalizing green infrastructure as essential public service provision. However, because of their classification as utility infrastructure, HOFOR is legally restricted from financing non-technical elements such as participatory design or biodiversity enhancements, limiting the potential for broader co-benefits. To overcome such silos, future NbS initiatives should align more closely with urban development and renewal agendas to enable more holistic, multifunctional outcomes.

Meanwhile in Glasgow, the Connecting Nature programme and its Open Space Strategy provide a strategic framework for mainstreaming NbS with economic sustainability. The Glasgow City Council has leveraged opportunity mapping, business model canvases and early-stage support for nature-based enterprises to explore self-sustaining financing models. The Open Space Strategy has already secured 500,000 GBP (670,000 USD) for pilot improvements, with plans to scale using Scottish Government funding supported by NbS spatial datasets. The Council has also partnered with NatureScot and Dark Matters Lab to develop TreesAI, an open-source platform valuing urban forest for carbon credit and ES

Kelvingrove cityscape in Glasgow, United Kingdom. The International Association of Horticultural Producers (AIPH).

markets. The initiative is structured to identify finance and enterprise models that help NbS become self-funded and economically scalable, embedding governance reforms that facilitate cross-departmental and community-led investments.

Nature-based enterprises that use NbS often find it difficult to secure the initial funding they require. Private investors are cautious, as they perceive NbS as risky and are uncertain about the financial returns. In this sense, a key bright spot for blended finance pathways emerges from the Connecting Nature Enterprise Platform which brings different financial actors including insurers, impact investors, and public funders, into the same conversation. Through initiatives like the NATURANCE dialogues, the platform has helped explore new ways to share risks and make NbS more appealing to private investors. For example, if public or donor support agree to cover the "first loss" in a project, commercial lenders and insurers become more willing to invest, knowing that part of their risk is reduced.

By connecting a marketplace of NbS suppliers and buyers with these finance discussions, the platform generates a consistent flow of real projects ready for investment. This approach goes beyond short-term grants and begins to demonstrate how NbS can be supported through ongoing, market-aligned funding. However, many current models mainly focus on technical risks, such as whether an NbS project will perform as planned. To realise their potential, future efforts should also consider broader benefits like community wellbeing, biodiversity, and urban regeneration. Doing so will help ensure that NbS are seen as essential, investable components of healthier, more resilient cities, rather than optional add-ons.

NbS Public Perceptions & Aspirations Bright Spots

For NbS to be socially acceptable and implemented equitably, the diversity of aspirations and needs must be fully acknowledged within decision-making processes. The climate adaptation and urban renewal project of Grønningen-Bispeparken in Copenhagen, Denmark is a good illuminating example of meaningful engagement of local residents and stakeholders in NbS decision-making. This project was developed in

The Grønningen-Bispeparken project integrates stormwater management with the resident's wishes for social meeting points and playscapes for children and youth. Oriol Garcia-Antúnez, 2025.

close dialogue with local residents and other stakeholders during the five years preceding its implementation phase. Between 2016 and 2021, social workers, anthropologists, and geographers were tasked with mapping and identifying the wishes and preferences of local residents and stakeholders. They employed various methods and activities, including hearings, online meetings, and interactive tools such as sandboxes and 3D models, to co-develop visions for the area. Moreover, the involvement of local artist Kerstin Bergendal was particularly important. Kerstin created sixteen short portrait films of area residents, which were integrated into the design process of the park. These ideas collected during this five-year preparatory phase formed the backbone of the requirements that landscape architects were asked to fulfill when submitting proposals during the project's competitive bidding process. This case illustrates that meaningful and inclusive engagement requires long-term investment; it is therefore crucial to allocate sufficient time and resources to foster trust, enable participation, and ensure that stakeholder perspectives genuinely shape NbS outcomes.

Governance & institutional bright spots

Fragmented institutional structures and siloed governance often hinder the effective implementation of NbS, especially in complex urban environments. However, two exemplary cases, the SPARK project in Aarhus, Denmark and the CLEVER Cities initiative in Milan, Italy, illustrate how mosaic governance can serve as a catalyst for institutional innovation and cross-sectoral collaboration in implementing NbS.

In Aarhus, SPARK brings together stormwater management, recreation, and public health through coordinated action among local authorities, utilities, and community actors. Through stakeholder workshops involving local institutions (such as Aarhus University, Aarhus Vand utilities, and municipal departments) the project highlighted how NbS can deliver dual ecosystem and public health benefits, while also

The CLEVER Cities program in South Thamesmead, London. Chiara Parretta, PLANES lab, University of Trento, n.d.

underlining the need for regulatory and collaborative frameworks that support early stakeholder engagement.

In Milan, the CLEVERCities project focused on revitalizing the underserved neighborhoods of Lorenteggio and Giambellino and the Tibaldi Station using NbS as a platform for social inclusion, resilience, and urban regeneration. What distinguished this initiative is its emphasis on co-creation: the municipality, community organizations, academic institutions, and private sector partners engaged in a participatory design process to plan and implement urban greening interventions. As such, research on the CLEVERCities governance typology observed a shift from top-down approaches to more collaborative, shared governance models throughout the project phases, with living labs serving as "nodes" to strengthen institutional capacity and enable more responsive planning structures

This model of mosaic governance enables deep institutional learning, as different actors bring diverse forms of knowledge (technical, local, experiential) into decision-making spaces. The governance process includes not only formal public bodies but also non-traditional actors such as social cooperatives and neighborhood councils, expanding the institutional ecosystem around NbS. Moreover, the CLEVERCities project has led to the creation of new procedural tools (e.g., co-design guidelines) that embed flexibility and responsiveness into the city's planning frameworks.

Both cases show how breaking down silos and fostering cooperation across governance levels and sectors can deliver multifunctional urban benefits. They highlight the institutional opportunity to embed flexibility, inclusivity, and responsiveness into planning systems as key ingredients for scaling and sustaining NbS through mosaic governance.

SPARK project implemented in Aarhus, Denmark. Kristine Jensen, 2022.

4 Recommendations and Future Directions

The Bosconavigli residential complex in Milan, Italy. Stefano Boeri Architetti, 2025.

The European Commission has recognized the value of NbS through several policy initiatives, including the EU Biodiversity Strategy for 2030 and the recently proposed Nature Restoration Law. These policies underscore the importance of NbS in addressing biodiversity loss, climate change, and urban sustainability, aligning with broader frameworks like the European Green Deal and Horizon Europe research programs.

Despite these strong policy frameworks, NbS implementation remains challenging, primarily due to barriers stemming from European urban environments' unique technical, economic, social, and governance dynamics.

This section outlines Key Recommendations and future directions for advancing NbS in European urban contexts. By addressing the Key Challenges – summarized in Section 3 into technical and knowledge barriers, economic and financial constraints, public perceptions of and aspirations for urban NbS, and governance and institutional barriers – these recommendations provide a practical roadmap for cities to enhance NbS uptake, secure long-term success, and align local efforts with existing EU policy goals. Key recommendations are also linked to representative case studies to illustrate how recommendations can be implemented in practice.

4.1 Technical and Knowledge Barriers

Many cities struggle to turn the growing conceptual knowledge of NbS into practical, coordinated action. We outline four key recommendations to help address challenges such as unclear definitions, integration with existing infrastructure, lack of technical standards, and insufficient long-term monitoring continue to hinder effective and scalable implementation.

KEY RECOMMENDATION 1:

Comprehensive approaches that provide implementing bodies with necessary design, planning, installation and maintenance skills are necessary to establish effective and impactful urban NbS at the city level.

Cities need integrating strategies that equip the relevant actors in NbS projects with the capacities for implementing them at different stages. A good example is the CLEVER Cities case study in Milan. The project had a strong focus on providing support and guidance to companies, designers, condominiums, and building owners interested in developing green roofs. This includes mapping potential areas for intervention in Milan and its metropolitan region, as well as a directory of companies and organizations offering relevant services and products.

The CLEVER Cities team providing support to local building owners in Milan. CleverCities, 2025.

KEY RECOMMENDATION 2:

Cities should allocate dedicated funding to monitor the ecological, social, and economic benefits of NbS over time, allowing these systems to adapt to specific socio-environmental contexts, while capturing data to evaluate NbS effectiveness.

Localized, context-sensitive monitoring provides a more accurate assessment of NbS performance and strengthens the case for ongoing support and investment. Integrating existing biodiversity assessment tools into NbS is a cost-effective approach. The CLEARING HOUSE project in Gelsenkirchen, Germany, shows how tracking data over time can help turn NbS knowledge into practical action. By setting up two forest plots in former industrial sites, the project helps measure ecosystem benefits and supports better planning for urban nature solutions. Another example is the Connecting Nature Glasgow case study, where a city scale impact assessment was created into a central resource that made it possible to support a more integrated and targeted decision-making process for NbS. This kind of databases also allow the promotion of data sharing and interdepartmental working.

CLEARING HOUSE project. Ian Whitehead.

Connecting Nature Glasgow. The International Association of Horticultural Producers (AIPH).

KEY RECOMMENDATION 3:

Biodiversity should be at the heart of NbS planning, design, and especially monitoring, to ensure that key ecological processes and species interactions are properly documented and supported.

Urban areas are often wrongly seen as having little value for biodiversity, which has led to underinvestment in biodiversity-focused NbS and weak monitoring practices (Spotswood et al., 2021). This gap is evident in many case studies and is made worse by the longer time and larger areas needed to see real change in urban biodiversity. Ignoring biodiversity risks creating designs that don't support ecosystem resilience or deliver long-term benefits to people (Knapp and MacIvor, 2023). It also misses the chance to strengthen people's connection with nature—an important driver of change in cities (O'Brien et al., 2025). This report calls for going beyond basic species counts, and instead using more holistic and practical frameworks to guide biodiversity efforts in urban NbS (Soga and Gaston, 2022). A good example of a framework for biodiversity benchmarking and monitoring is the Urban Biodiversity Score (Bynatur app.).

KEY RECOMMENDATION 4:

Prioritize support for small, community-driven NbS initiatives as essential building blocks for scalable and inclusive urban transformation.

Smaller, localized NbS projects, often initiated by active citizens, offer substantial value in terms of community engagement and local impact. Projects like the Rumon case demonstrate the potential for small-scale initiatives to generate valuable knowledge and insights that can be scaled up or adapted for wider application. Supporting these community-driven projects not only fosters a sense of ownership but also enriches the broader knowledge base of NbS by capturing the experiences and successes of diverse, grassroots approaches. As cities and organizations seek to scale NbS, investing in these smaller initiatives can provide essential learnings that inform and enhance larger projects.

Rumon association growing a food forest in Finale Emilia. Antonio Luizo and Elisa Debbi, 2025.

4.2 Economic and Financial Constraints

Perceived financial risks and unclear cost-benefit analyses make it difficult to secure sustained investment in NbS. Without strong evidence of long-term value and equitable benefit distribution, NbS often lose out to conventional infrastructure in funding decisions.

KEY RECOMMENDATION 5:

Cities should establish stable, long-term funding mechanisms – such as public-private partnerships – to ensure ongoing management and maintenance of NbS.

For example, the Connecting Nature project, which in this report comprises two analyzed case studies, created the Connecting Nature Enterprise Platform in response to the need of having Nature-Based Enterprises dedicated to working for and with nature. Currently, the platform also hosts a growing resource library filled with knowledge and insights on new technologies and opportunities within the nature-positive economy. These types of platforms also encourage a broader private sector participation through incentives and risk-sharing models to boost the adoption of NbS.

Logo of the Connecting Nature Enterprise Platform.

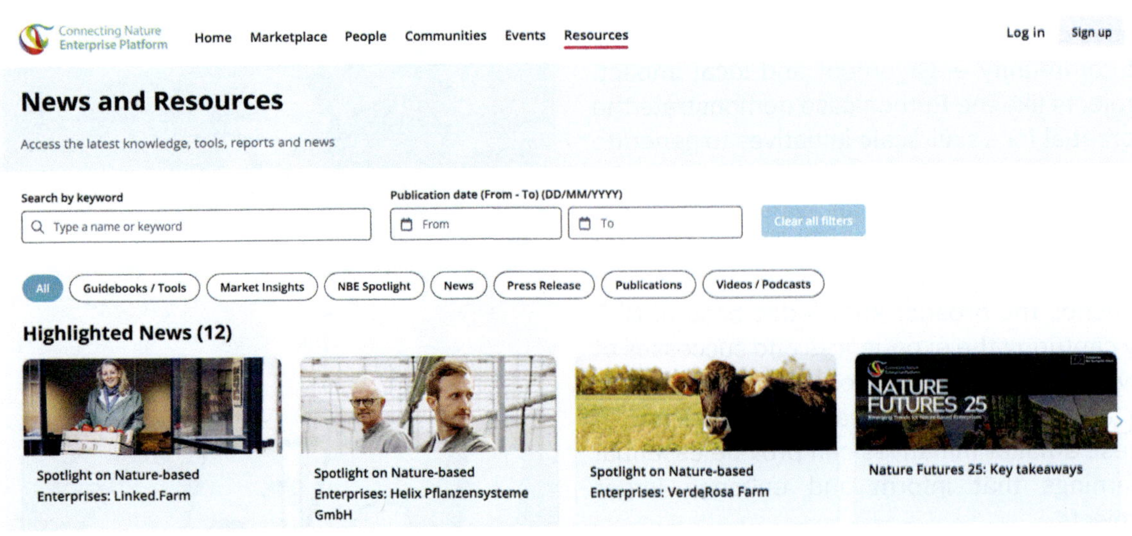

A screenshot of the "Resources" section on the website, which features news, blog posts, reports, website recaps, and more. Taken on March 12, 2026.

KEY RECOMMENDATION 6:

Local governments should use funding models that support community involvement to help maintain NbS over time.

Investment in stewardship – understood as funding models that empower local ownership and long-term care – can work around financial constraints to encourage active, long-term involvement from citizens and local actors. For example, Tampere's UNaLab project showed that integrating innovation vouchers into the NbS funding model can successfully foster this type of community engagement. By providing innovation vouchers, the city enabled residents and local actors, beyond just municipal staff, to develop and implement novel NbS ideas, making them active contributors to the city's sustainability goals. While the Vuores Living Lab applied this approach in a new development area, the project findings suggest that targeting similar funding to older, denser urban neighborhoods – where green space and resources are often limited – could further enhance the reach and impact of NbS efforts.

UNaLab in Tampere, Finland. Eugenia Castellazzi, 2025.

4.3 Public Perceptions of and Aspirations for Urban Nature-based Solutions

Public and political acceptance of NbS remains fragile, shaped by limited awareness, mistrust in their effectiveness, and competing demands for urban space. When NbS do not reflect local values or aspirations, they risk low support, which in turn undermines long-term political commitment.

KEY RECOMMENDATION 7:

A strong foundation of public trust starts with involving diverse communities in NbS planning and using inclusive outreach methods.

Engaging residents early and meaningfully helps ensure that NbS reflect local needs, gain public support, and deliver both social and ecological benefits. The Gelsenkirchen case shows how listening to a wide range of community perspectives led to strategic goals for a post-industrial forest that were not only environmentally effective but also locally accepted. In Zagreb, co-creation efforts specifically included outreach to people with disabilities and local associations, helping to design NbS spaces and activities that were inclusive and widely supported.

KEY RECOMMENDATION 8:

Ensure that co-creation is at the heart of local NbS projects with transparent, inclusive, and open communication and participatory processes.

This means sharing information clearly so that everyone involved in implementing NbS understands the goals and reasons behind co-created solutions. This helps build public trust and keeps projects accountable. To make co-creation more credible and effective, it's important to define who should be involved – such as local communities, researchers, and businesses – through stakeholder mapping, inclusive governance, and better communication of NbS benefits. A good example of successful co-creation is the recent climate change adaptation and urban renewal project of Grønningen-Bispepaken,Copenhagen, Denmark. The expectations and aspirations of members of the local community – elicited by means of participatory and artistic workshops – became the backbone and requirements to be fulfilled by the landscape architects submitting proposals as part of the project's competitive bidding process.

Grønningen-Bispepaken, Copenhagen, Denmark. Oriol Garcia-Antúnez, 2025.

4.4 Governance and Institutional Barriers

Fragmented responsibilities, regulatory gaps, and siloed decision-making often delay or derail NbS projects. Without integrated governance and supportive policy frameworks, cities struggle to align stakeholders and sustain momentum for transformative change.

KEY RECOMMENDATION 9:

Establish governance frameworks to coordinate across departments, reducing siloed decision-making and ensuring integrated NbS approaches.

The SPARK case study in Aarhus, Denmark, represents a good attempt to integrate multiple policy domains, such as physical and mental health, outdoor recreation, stormwater management, and biodiversity. Similarly, the CLEVER Cities project in Milan demonstrated how NbS can be embedded within complex, multi-level governance systems. Milan's approach involves aligning urban regeneration efforts with environmental goals through internal co-design processes within public administrations. By fostering collaboration across municipal departments and engaging citizens and stakeholders, the project helped break down both horizontal and vertical silos. This co-production model not only integrated diverse urban agendas (including climate adaptation, social inclusion, and biodiversity) but also built long-term institutional capacity that remained even after the project conclusion.

SPARK project implemented in Aarhus, Denmark. Kristine Jensen, 2022.

KEY RECOMMENDATION 10:

Clear policies and targets are needed to scale up effective and accountable NbS.

Update policies to provide clear regulatory support for NbS, including guidelines for procurement, risk management, and performance standards. Establish binding targets and metrics for NbS effectiveness. For example in A Coruña, the implementation of community gardens has been favored by the implementation of an integrated strategy for sustainable urban development (EIDUS) whose implementation is supported by ERDF funds, and is one of the pioneer cities in Spain to adopt the local implementation of the new Spanish Urban Agenda, which is aligned with the European Urban Agenda and the UN Sustainable Development Goals.

Connecting Nature project in A Coruña. Connecting Nature, 2019.

KEY RECOMMENDATION 11:

Ensure that political and financial support for NbS projects is protected from shifts in leadership by embedding NbS goals within broader city or national framework.

For example, leveraging the 24-month window for developing National Restoration Plans to include specific actions that promote NbS co-design and implementation is possible. This can guide member states in creating tailored NbS approaches that align with national priorities and support biodiversity and ecosystem health. Moreover, integrating NbS into Restoration Plans offers member states a structured opportunity to innovate in NbS planning and ensure consistent standards, aligning local actions with broader EU goals. Additionally, incorporating NbS within Urban Greening Plans can further strengthen long-term commitments, ensuring alignment with the EU Biodiversity Strategy for 2030.

The European Parliament voting on the Nature Restoration Law in 2023. European Union, 2023.

4.5 Limitations and Future Research Pathways

This report provides a snapshot of NbS implementation across European cities, but several limitations should be acknowledged to guide future work. A key limitation is the concentration of sources from EU-funded projects and publications, which may not fully reflect the diversity of NbS practices across Europe. Important lessons from earlier approaches, such as ecological restoration or ecological engineering under LIFE and INTERREG programmes, may have been overlooked. Additionally, the focus on public or semi-public land excluded initiatives on privately owned land, which play a vital role in scaling NbS out. Future research should explore governance models and incentives to support implementation on private properties.

The linguistic diversity of Europe meant that some locally produced knowledge – particularly grey literature – was underrepresented. Despite prioritizing multilingual sourcing where possible, language barriers remain a significant obstacle to inclusive knowledge-sharing

Establishing a central repository of case studies, data, and best practices would facilitate peer learning and support cities in refining their NbS strategies.

Data availability also posed challenges. Some grassroots and bottom-up initiatives were difficult to include due to outdated or inaccessible project websites and vague descriptions lacking the detail needed for replication. Moreover, the linguistic diversity of Europe meant that some locally produced knowledge – particularly grey literature – was underrepresented. Despite prioritizing multilingual sourcing where possible, language barriers remain a significant obstacle to inclusive knowledge-sharing (Dushkova et al., 2025). This underlines the need for future research to emphasize multilingual collaboration and open-access dissemination.

As cities increasingly adopt NbS, ongoing research is essential to ensure these solutions are effective, scalable, and integrated into broader policy frameworks. Many projects are monitored only for the duration of initial funding periods, limiting long-term impact and learning. Scaling up NbS at national and EU levels will require robust protocols that match the scale and characteristics of each intervention. Technologies such as Geographic Information Systems (GIS) and remote sensing can help provide consistent, real-time data to support adaptive planning.

Current methods for tracking ecosystem services often fail to capture the complexity of urban NbS. Large-scale assessments exist, but more localized, tailored approaches are needed to reflect the multiple co-benefits NbS can deliver – such as cooling cities, improving air quality, enhancing biodiversity, and strengthening social resilience. These benefits, however, may interact in complex ways, sometimes creating trade-offs. For instance, a flood management solution may reduce space for recreation, or maximizing vegetation for cooling might not align with biodiversity goals. More research is needed to understand and manage these trade-offs, helping cities design NbS that maximize synergies and balance competing priorities.

Finally, sharing knowledge across cities remains critical. Establishing a central repository of case studies, data, and best practices would facilitate peer learning and support cities in refining their NbS strategies. Strengthening research in these areas will help NbS reach their full potential in creating greener, more resilient urban spaces.

5 | References

1. Akoumianaki, I. and Pakeman, R. (2023) How Does Biodiversity Contribute to Nature-based Solutions? Available at: https://www.biodiversa.eu/2024/08/06/how-does-biodiversity-contribute-to-nature-based-solutions/ (Accessed: October 30, 2024).

2. Albert, C. et al. (2019) "Addressing Societal Challenges through Nature-Based Solutions: How Can Landscape Planning and Governance Research Contribute?," Landscape and Urban Planning, 182, pp. 12–21. Available at: https://doi.org/10.1016/j.landurbplan.2018.10.003.

3. Andersson, E. et al. (2019) "Enabling Green and Blue Infrastructure to Improve Contributions to Human Well-Being and Equity in Urban Systems," BioScience, 69(7), pp. 566–574. Available at: https://doi.org/10.1093/biosci/biz058.

4. Andersson, E. et al. (2021) "A Context-Sensitive Systems Approach for Understanding and Enabling Ecosystem Service Realization in Cities," Ecology and Society, 26(2). Available at: https://doi.org/10.5751/ES-12411-260235.

5. Anguelovski, I. and Corbera, E. (2023) "Integrating Justice in Nature-Based Solutions to Avoid Nature-Enabled Dispossession," Ambio, 52(1), pp. 45–53. Available at: https://doi.org/10.1007/s13280-022-01771-7.

6. Baud, I. et al. (2021) "The Urban Governance Configuration: A Conceptual Framework for Understanding Complexity and Enhancing Transitions to Greater Sustainability in Cities," Geography Compass, 15(5), p. e12562. Available at: https://doi.org/10.1111/gec3.12562.

7. Carayannis, E.G., Barth, T.D. and Campbell, D.F. (2012) "The Quintuple Helix Innovation Model: Global Warming as a Challenge and Driver for Innovation," Journal of Innovation and Entrepreneurship, 1(1), p. 2. Available at: https://doi.org/10.1186/2192-5372-1-2.

8. Castellazzi, E., Hakkarainen, V. and Raymond, C.M. (2024) "Role-Play as a Catalyst for Justice Awareness: A Relational Approach to Nature-Based Solutions with Secondary School Students," Ecosystems and People, 20(1), p. 2430591. Available at: https://doi.org/10.1080/26395916.2024.2430591.

9. City of Copenhagen (2012) Copenhagen Climate Adaptation Plan. Available at: https://international.kk.dk/sites/default/files/2021-09/Copenhagen%20Climate%20Adaptation%20Plan%20-%202011.pdf.

10. Clar, C., Prutsch, A. and Steurer, R. (2013) "Barriers and Guidelines for Public Policies on Climate Change Adaptation: A Missed Opportunity of Scientific Knowledge-Brokerage," Natural Resources Forum, 37(1), pp. 1–18. Available at: https://doi.org/10.1111/1477-8947.12013.

11. Clever Cities (2021) Co-Design of Nature-Based Solutions with Citizens: Learnings from a Sector Literature Review. Available at: https://clevercities.eu/fileadmin/user_upload/Resources/Co-design_NBS_Learnings_from_a_sector_literature_review.pdf.

12. Cohen-Shacham, E. et al. (2019) "Core Principles for Successfully Implementing and Upscaling Nature-based Solutions," Environmental Science & Policy, 98, pp. 20–29. Available at: https://doi.org/10.1016/j.envsci.2019.04.014.

13. Cortinovis, C., Geneletti, D. and Haase, D. (2022) "Higher immigration and lower land take rates are driving a new densification wave in European cities," npj Urban Sustainability, 2(1), p. 19. Available at: https://doi.org/10.1038/s42949-022-00062-0.

14. Demuzere, M. et al. (2014) "Mitigating and Adapting to Climate Change: Multi-functional and Multi-Scale Assessment of Green Urban Infrastructure," Journal of Environmental Management, 146, pp. 107–115. Available at: https://doi.org/10.1016/j.jenvman.2014.07.025.

15. Dushkova, D. et al. (2024) "Urban Ecosystem Services Research in Russia: Systematic Review on the State of the Art," Ambio [Preprint]. Available at: https://doi.org/10.1007/s13280-024-02102-8.

16. Dushkova, D. et al. (2025) "Urban ecosystem services research in Russia: Systematic review on the state of the art," Ambio, 54(4), pp. 577–602. Available at: https://doi.org/10.1007/s13280-024-02102-8.

17. Dushkova, D. and Haase, D. (2023) "Resilient Cities, Healthy Communities, and Sustainable Future: How Do Nature-Based Solutions Contribute?," in P. Liamputtong (ed.) Handbook of Social Sciences and Global Public Health. Cham: Springer International Publishing, pp. 2031–2055. Available at: https://doi.org/10.1007/978-3-031-25110-8_133.

18. Eggermont, H. et al. (2015) "Nature-Based Solutions: New Influence for Environmental Management and Research in Europe," GAIA—Ecological Perspectives for Science and Society, 24(4), pp. 243–248. Available at: https://doi.org/10.14512/gaia.24.4.9.

19. EL Harrak, M. and Lemaitre, F. (2023) European Roadmap to 2030 for NetworkNatureResearch and Innovation on Nature-based Solutions.

20. European Commission (2013) Open Innovation 2013. Publications Office of the European Union. Available at: https://data.europa.eu/doi/10.2759/87245 (Accessed: May 23, 2024).

21. European Commission (2015) Towards an EU Research and Innovation Policy Agenda for Nature-Based Solutions & Re-Naturing Cities: Final Report of the Horizon 2020 Expert Group on 'Nature Based Solutions and Re Naturing Cities' : (Full Version). Publications Office of the European Union. Available at: https://data.europa.eu/doi/10.2777/765301 (Accessed: April 15, 2024).

22. European Commission (2020) Nature-Based Solutions: State of the Art in EU-funded Projects—European Commission. Available at: https://research-and-innovation.ec.europa.eu/knowledge-publications-tools-and-data/publications/all-publications/nature-based-solutions-state-art-eu-funded-projects_en (Accessed: January 20, 2025).

23. European Commission et al. (2023) Guidelines for Co-Creation and Co-Governance of Nature-Based Solutions: Insights Form EU Funded Projects. Edited by G. Lupp, I. Mahmoud, and I. Ferreira. Publications Office of the European Union. Available at: https://data.europa.eu/doi/10.2777/157060 (Accessed: May 23, 2024).

24. European Commission (2023a) Nature-Based Solutions: EU Funded Nbs Research Projects Tackle the Climate and Biodiversity Crisis. Publications Office of the European Union. Available at: https://data.europa.eu/doi/10.2848/879543 (Accessed: October 31, 2024).

25. European Commission (2023b) Nature-Based Solutions for Biodiversity and Climate. Available at: https://knowledge4policy.ec.europa.eu/biodiversity/nature-based-solutions-biodiversity-climate_en#:~:text=The%20umbrella%20concept%20of%20NBS,drainage%20systems%2C%20and%20ecological%20engineering.

26. Evers, D., Katurić, I. and Van Der Wouden, R. (2024) "Urbanization in Europe 2000–2018: Amount, Density, and Form," in Evers, D., Katurić, I., and Van Der Wouden, R., Urbanization in Europe. Cham: Springer International Publishing (Sustainable Urban Futures), pp. 15–52. Available at: https://doi.org/10.1007/978-3-031-62261-8_2.

27. Faivre, N. et al. (2017) "Nature-Based Solutions in the EU: Innovating with nature to address social, economic and environmental challenges," Environmental Research, 159, pp. 509–518. Available at: https://doi.org/10.1016/j.envres.2017.08.032.

28. Frantzeskaki, N. and Kabisch, N. (2016) "Designing a Knowledge Co-Production Operating Space for Urban Environmental Governance—Lessons from Rotterdam, Netherlands and Berlin, Germany," Environmental Science & Policy, 62, pp. 90–98. Available at: https://doi.org/10.1016/j.envsci.2016.01.010.

29. Garcia Antúnez, O. (2024) Unpacking and Fostering Socially Acceptable and Just Urban Green Space Transformations. Department of Geosciences and Natural Resource Management, Faculty of Science, University of Copenhagen.

30. Garcia-Antúnez, O. et al. (2023) "Unpacking Public Perceptions of Carbon Sequestration and Storage in Urban Greenery: Implications for the Social Acceptability of Carbon-Oriented Nature-Based Solutions," Nature-Based Solutions, 4, p. 100087. Available at: https://doi.org/10.1016/j.nbsj.2023.100087.

31. Grilo, F. et al. (2020) "Using Green to Cool the Grey: Modelling the Cooling Effect of Green Spaces with a High Spatial Resolution," Science of The Total Environment, 724, p. 138182. Available at: https://doi.org/10.1016/j.scitotenv.2020.138182.

32. Haaland, C. and Van Den Bosch, C.K. (2015) "Challenges and strategies for urban green-space planning in cities undergoing densification: A review," Urban Forestry & Urban Greening, 14(4), pp. 760–771. Available at: https://doi.org/10.1016/j.ufug.2015.07.009.

33. Haase, A. (2017) "The Contribution of Nature-Based Solutions to Socially Inclusive Urban Development–Some Reflections from a Social-environmental Perspective," in N. Kabisch et al. (eds.) Nature-Based Solutions to Climate Change Adaptation in Urban Areas: Linkages between Science, Policy and Practice. Cham: Springer International Publishing, pp. 221–236. Available at: https://doi.org/10.1007/978-3-319-56091-5_13.

34. Hering, D. et al. (2023) "Securing success for the Nature Restoration Law," Science, 382(6676), pp. 1248–1250. Available at: https://doi.org/10.1126/science.adk1658.

35. Hölscher, K. et al. (2023) "Strategies for mainstreaming nature-based solutions in urban governance capacities in ten European cities,"

npj Urban Sustainability, 3(1), p. 54. Available at: https://doi.org/10.1038/s42949-023-00134-9.

36. Hölscher, K. et al. (2024) "Embedding Co-Production of Nature-Based Solutions in Urban Governance: Emerging Co-Production Capacities in Three European Cities," Environmental Science & Policy, 152, p. 103652. Available at: https://doi.org/10.1016/j.envsci.2023.103652.

37. Hudson, G., Hart, S. and Verbeek, A. (2023) Investing in Nature-Based Solutions State-of-play and Way Forward for Public and Private Financial Measures in Europe. European Investment Bank.

38. International Union for Conservation of Nature (2020) Ensuring Effective Nature-based Solutions. Available at: https://www.iucn.org/resources/issues-brief/ensuring-effective-nature-based-solutions (Accessed: June 7, 2024).

39. Kabisch, N. et al. (2016) "Nature-Based Solutions to Climate Change Mitigation and Adaptation in Urban Areas: Perspectives on Indicators, Knowledge Gaps, Barriers, and Opportunities for Action," Ecology and Society, 21(2). Available at: https://doi.org/10.5751/ES-08373-210239.

40. Kato-Huerta, J. and Geneletti, D. (2022) "Environmental justice implications of nature-based solutions in urban areas: A systematic review of approaches, indicators, and outcomes," Environmental Science & Policy, 138, pp. 122–133. Available at: https://doi.org/10.1016/j.envsci.2022.07.034.

41. Kato-Huerta, J. and Geneletti, D. (2023) "Analysing the treatment of environmental justice and nature-based solutions in the Urban Climate Action Plans of Latin American metropolitan areas," Local Environment, 28(11), pp. 1388–1409. Available at: https://doi.org/10.1080/13549839.2023.2221431.

42. Knapp, S. and MacIvor, J.S. (2023) "Chapter 5: Nature-based Solutions and Biodiversity: Synergies, Trade-Offs, and Ways Forward," in. Available at: https://www.elgaronline.com/edcollchap-oa/book/9781800376762/book-part-9781800376762-14.xml (Accessed: February 3, 2025).

43. Kolokotsa, D. et al. (2020) "On the Impact of Nature-Based Solutions on Citizens' Health & Well Being," Energy and Buildings, 229, p. 110527. Available at: https://doi.org/10.1016/j.enbuild.2020.110527.

44. Lampinen, J. et al. (2021) "Acceptance of Near-Natural Greenspace Management Relates to Ecological and Socio-Cultural Assigned Values among European Urbanites," Basic and Applied Ecology, 50, pp. 119–131. Available at: https://doi.org/10.1016/j.baae.2020.10.006.

45. Lampinen, J. et al. (2023) "Mapping Public Support for Urban Green Infrastructure Policies across the Biodiversity-Climate-Society -Nexus," Landscape and Urban Planning, 239, p. 104856. Available at: https://doi.org/10.1016/j.landurbplan.2023.104856.

46. Langemeyer, J. et al. (2022) The INTERLACE—Pilot Assessment Framework for Restorative Nature-based Solutions: A Stepwise, Modular and Hierarchical Approach. Deliverable 3.2. INTERLACE project.

47. Li, M. et al. (2025) "Solution to What? Global Assessment of Nature-Based Solutions, Urban Challenges, and Outcomes," Landscape and Urban Planning, 256, p. 105294. Available at: https://doi.org/10.1016/j.landurbplan.2025.105294.

48. Linnerooth-Bayer, J. et al. (2023) Learning from NBS implementation barriers. International Institute of Applied Systems Analysis. Available at: https://pure.iiasa.ac.at/id/eprint/19283/1/deliverable-d5-4.pdf.

49. Lorenzoni, I., Nicholson-Cole, S. and Whitmarsh, L. (2007) "Barriers Perceived to Engaging with Climate Change among the UK Public and Their Policy Implications," Global Environmental Change, 17(3), pp. 445–459. Available at: https://doi.org/10.1016/j.gloenvcha.2007.01.004.

50. Maes, J. et al. (2020) Mapping and Assessment of Ecosystems and their Services: An EU ecosystem assessment. Luxembourg: Publications Office of the European Union. Available at: https://data.europa.eu/doi/10.2760/757183 (Accessed: October 18, 2025).

51. Maurer, M. et al. (2023) "A Social-Ecological-Technological System Approach to Just Nature-Based Solutions: A Case of Digital Participatory Mapping of Meaningful Places in a Marginalized Neighborhood in Copenhagen, Denmark," Urban Forestry & Urban Greening, 89, p. 128120. Available at: https://doi.org/10.1016/j.ufug.2023.128120.

52. McPhearson, T. et al. (2016) "Advancing Understanding of the Complex Nature of Urban Systems," Ecological Indicators, 70, pp. 566–573. Available at: https://doi.org/10.1016/j.ecolind.2016.03.054.

53. Nesshöver, C. et al. (2017) "The Science, Policy and Practice of Nature-Based Solutions: An Interdisciplinary Perspective," Science of The Total Environment, 579, pp. 1215–1227. Available at: https://doi.org/10.1016/j.scitotenv.2016.11.106.

54. O'Brien, K. et al. (2025) IPBES Transformative Change Assessment: Summary for Policymakers. Zenodo. Available at: https://doi.org/10.5281/ZENODO.11382230.

55. Raymond, C.M., Frantzeskaki, N., et al. (2017) "A Framework for Assessing and Implementing the Co-Benefits of Nature-Based Solutions in Urban Areas," Environmental Science & Policy, 77, pp. 15–24. Available at: https://doi.org/10.1016/j.envsci.2017.07.008.

56. Raymond, C.M., Breil, M., et al. (2017) An Impact Evaluation Framework to Support Planning and Evaluation of Nature-Based Solutions Projects. Report Prepared by the EKLIPSE Expert Working Group on Nature-based Solutions to Promote Climate Resilience in Urban Areas. Centre for Ecology and Hydrology. Available at: https://ora.ox.ac.uk/objects/uuid:3ecfc907-1971-473a-87f3-63d1204120f0 (Accessed: November 29, 2024).

57. Russo, A. and Cirella, G.T. (2021) "Urban Ecosystem Services: Current Knowledge, Gaps, and Future Research," Land, 10(8), p. 811. Available at: https://doi.org/10.3390/land10080811.

58. Sarabi, S. et al. (2020) "Uptake and implementation of Nature-Based Solutions: An analysis of barriers using Interpretive Structural Modeling," Journal of Environmental Management, 270, p. 110749. Available at: https://doi.org/10.1016/j.jenvman.2020.110749.

59. Sari, R. et al. (2023) "Improving the Climate Resilience of European Cities via Socially Acceptable Nature-Based Solutions," npj Urban Sustainability, 3(1), pp. 1–13. Available at: https://doi.org/10.1038/s42949-023-00090-4.

60. Schmalzbauer, A. (2018) "Barriers and Success Factors for Effectively Cocreating Nature-Based Solutions for Urban Regeneration. Deliverable 1.1.1, CLEVER Cities, H2020 Grant No. 776604."

61. Schröter, B. et al. (2022) "Beyond Demonstrators—tackling fundamental problems in amplifying nature-based solutions for the post-COVID-19 world," npj Urban Sustainability, 2(1), p. 4. Available at: https://doi.org/10.1038/s42949-022-00047-z.

62. Seddon, N. et al. (2020) "Understanding the Value and Limits of Nature-Based Solutions to Climate Change and Other Global Challenges," Philosophical Transactions of the Royal Society B: Biological Sciences, 375(1794), p. 20190120. Available at: https://doi.org/10.1098/rstb.2019.0120.

63. Soga, M. and Gaston, K.J. (2022) "Towards a Unified Understanding of Human–Nature Interactions," Nature Sustainability, 5(5), pp. 374–383. Available at: https://doi.org/10.1038/s41893-021-00818-z.

64. Spotswood, E.N. et al. (2021) "The Biological Deserts Fallacy: Cities in Their Landscapes Contribute More than We Think to Regional Biodiversity," BioScience, 71(2), pp. 148–160. Available at: https://doi.org/10.1093/biosci/biaa155.

65. Tunstall, S.M. et al. (2000) "River Restoration: Public Attitudes and Expectations," Water and Environment Journal, 14(5), pp. 363–370. Available at: https://doi.org/10.1111/j.1747-6593.2000.tb00274.x.

66. UNaLAB Toolkit (2024). Available at: https://unalab.enoll.org/ (Accessed: August 27, 2024).

67. Vignola, R. et al. (2009) "Ecosystem-Based Adaptation to Climate Change: What Role for Policy-Makers, Society and Scientists?," Mitigation and Adaptation Strategies for Global Change, 14(8), pp. 691–696. Available at: https://doi.org/10.1007/s11027-009-9193-6.

68. Vujcic, M. et al. (2017) "Nature based solution for improving mental health and well-being in urban areas," Environmental Research, 158, pp. 385–392. Available at: https://doi.org/10.1016/j.envres.2017.06.030.

69. Wamsler, C. et al. (2020) "Environmental and Climate Policy Integration: Targeted Strategies for Overcoming Barriers to Nature-Based Solutions and Climate Change Adaptation," Journal of Cleaner Production, 247, p. 119154. Available at: https://doi.org/10.1016/j.jclepro.2019.119154.

70. Wilk, B. et al. (2020) Guidelines for Co-Designing and Co-Implementing Green Infrastructure in Urban Regeneration Processes. D 2.10, proGIreg. Horizon 2020 Grant Agreement No 776528. European Commission.

71. Wolff, M. and Haase, D. (2019) "Mediating Sustainability and Liveability—Turning Points of Green Space Supply in European Cities," Frontiers in Environmental Science, 7, p. 61. Available at: https://doi.org/10.3389/fenvs.2019.00061.

72. Zwierzchowska, I. et al. (2019) "Introducing Nature-Based Solutions into Urban Policy – Facts and Gaps. Case Study of Poznań," Land Use Policy, 85, pp. 161–175. Available at: https://doi.org/10.1016/j.landusepol.2019.03.025.

Visual References

1. The redevelopment project of the Kursaal in Abano Terme, Italy. Stefano Boeri Architetti, 2025. https://www.stefanoboeriarchitetti.net/en/project/kursaal-abano-terme/
2. A Green Promenade for Cagliari, Italy. Stefano Boeri Architetti, 2025. https://www.stefanoboeriarchitetti.net/en/project/a-green-promenade-for-cagliari-seafront/
3. DANA in Valencia, Spain, October 2024. Manuel Pérez Garcia and Estefania Monerri Mínguez, 2024. https://commons.wikimedia.org/wiki/File:Cami%C3%B3_de_la_UME_(ex%C3%A8rcit)_buidant_d%27aigua_un_baix_en_Catarroja_DANA_2024.jpg
4. Storm Bodil in Denmark, December 2013. julochka, 2013. https://www.flickr.com/photos/julochka/11296719093/
5. Heatwave in Southern Europe, July 2023. The European Space Agency, 2025. https://www.esa.int/ESA_Multimedia/Images/2025/07/Southern_Europe_s_land_and_sea_sizzles
6. The Ca'delle Alzaie residential project in Treviso, Italy. Stefano Boeri Architetti, 2021. https://www.stefanoboeriarchitetti.net/project/ca-delle-alzaie/
7. The Wonderwoods vertical forest project in Utrecht, the Netherlands. Stefano Boeri Architetti, 2025. https://www.stefanoboeriarchitetti.net/project/wonderwoods-2/
8. Intergenerational participatory workshop in Granollers, where youth and elderly participants expressed their priorities, aspirations, and expectations for the CoCoNat 2025 project using a variety of participatory tools. The City of Granollers, 2025. www.granollers.cat/trobadaintergeneracional#fotos
9. More than two hundred companies, associations, and non-governmental organizations join to urge national leaders to reaffirm their dedication to the European Green Deal. European Council/European Union, 2024. https://balkangreenenergynews.com/firmsassociations-ngos-to-urge-eu-leaders-to-confirm-europeangreen-deal-after-european-elections/
10. Flowers blooming in a Hørgården garden. Lft3b, 2011.https://da.wikipedia.org/wiki/Urbanplanen#/media/Fil:Urbanplanen_12.jpg
11. View of the housing area from the garden. Selveje Netværket. https://www.selvejenetvaerket.dk/plejecentret-hoergaarden-2/
12. SPARK project implemented in Aarhus, Denmark. Kristine Jensen, 2022. https://www.kristinejensen.dk/projekter/spark-marselisborgcentret/
13. The Rheinelbe Industrial Forest in Germany. Astrid Hamm, 2006. https://www.researchgate.net/figure/Rheinelbe-Colliery-Succession-forestat-different-stages_fig9_27655438
14. Kelvingrove cityscape in Glasgow, United Kingdom. The International Association of Horticultural Producers (AIPH). https://aiph.org/green-citycase-studies/glasgow-scotland/
15. SPARK project implemented in Aarhus, Denmark. Kristine Jensen, 2022. https://www.kristinejensen.dk/projekter/spark-marselisborgcentret/
16. The Bosconavigli residential complex in Milan, Italy. Stefano Boeri Architetti, 2025. https://www.stefanoboeriarchitetti.net/project/bosconavigli/
17. The CLEVER Cities team providing support to local building owners in Milan. CleverCities, 2025. https://clevercities.eu/news/c-search-uid-2s7exgvj/
18. Connecting Nature Glasgow. The International Association of Horticultural Producers (AIPH). https://aiph.org/green-citycase-studies/glasgow-scotland/
19. Logo of the Connecting Nature Enterprise Platform. Connecting Nature Enterprise Platform, 2026. https://naturebasedenterprise.com/topics
20. The European Parliament voting on the Nature Restoration Law in 2023. European Union, 2023 https://www.euronews.com/my-europe/2023/07/12/nature-restoration-law-survives-knife-edge-vote-in-the-european-parliament-amid-right-wing

6 Appendices

A Details of the Methods Applied (Task 1 and Task 2)

This section outlines the common protocol used across all regional assessments of the Global Roadmap for Urban Nature-based Solutions, along with region-specific methodological choices. Together, these methods synthesize the academic literature on urban NbS and assess geographical, bibliometric, and thematic trends. The overall process is summarized in Figure 10.

NATURA Global Roadmap Process Diagram

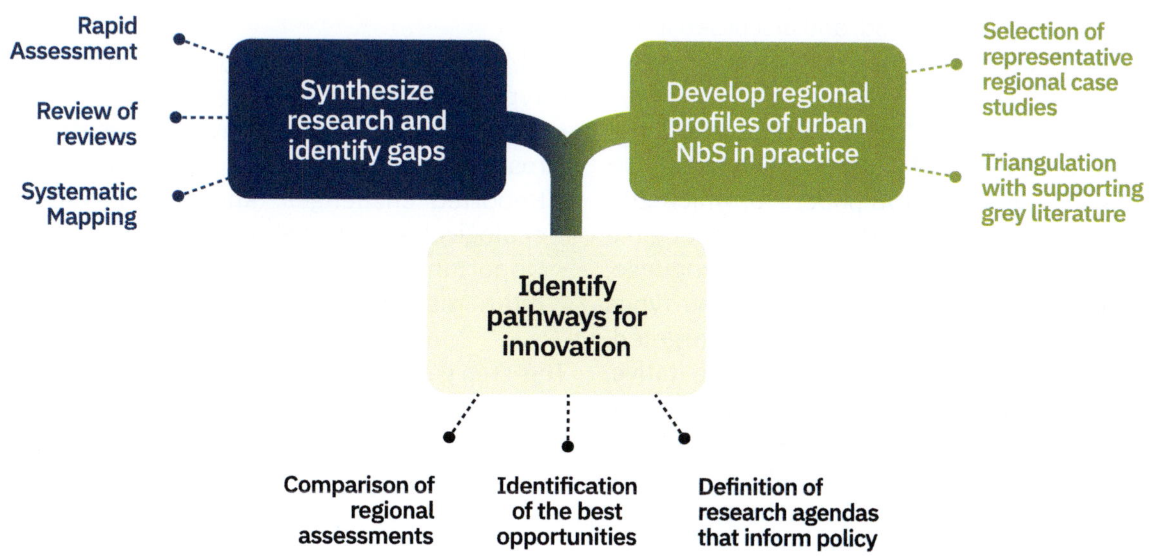

Figure 10: Overall Process of the Global Roadmap for Urban Nature-based Solutions

Task 1 | Literature Review

Rapid Assessment

Task 1.1. provides a high-level mapping of the academic landscape on nature-based solutions and related concepts. A SCOPUS search (36,123 papers) was conducted using a comprehensive set of NbS-related terms, restricted to English and urban contexts. Automated scripts in RStudio identified both regional focus (based on country and city keywords) and conceptual categories (based on NbS-related terms).

Validation checks showed 97% accuracy for region matching and 95.9% for concept matching. The analysis examined temporal publication trends, geographic distributions, and terminological overlaps to position NbS within the broader environmental literature and inform subsequent review and mapping tasks.

The terms explored included:

- blue infrastructure,
- blue-green infrastructure,
- disaster-risk reduction,
- eco-tourism,
- ecological engineering,
- ecological infrastructure,
- ecological restoration,
- ecosystem services,
- ecosystem-based adaptation,
- ecosystem restoration,
- ecosystem-based approaches,
- ecosystem-based management,
- green infrastructure,
- green lungs,
- green space management,
- green development,
- greenbelt,
- landscape ecology approach,
- linear parks,
- natural infrastructure,
- resilient infrastructure,
- sponge city,
- sustainable land management,
- sustainable urban drainage,
- traditional ecological knowledge ,
- urban forestry,
- urban greening,
- urban resilience, and
- water-sensitive urban design.

Review of Reviews

A systematic review of reviews was undertaken to synthesize how urban NbS are conceptualized, reviewed, and discussed in the academic literature.

A SCOPUS search (“nature-based solution*” AND “review” AND (“urban” OR “city” OR “cities”)) identified 450 papers, of which 203 met inclusion criteria after screening. Eligible papers were limited to English-language, urban-focused, and a range of review types (systematic, scoping, bibliometric, targeted, narrative, integrative, comparative, qualitative, critical, rapid, structured, and scientometric reviews), as well as systematic maps, quantitative systematic literature reviews, meta-analyses, bibliographic reviews, thematic analyses, and network analyses).

Each paper was fully screened and manually coded for:

- NbS concepts referenced (e.g., GI, ES, EbA)
- Type and scope of review
- Geographic and thematic trends, biases, and research gaps
- Reported challenges (knowledge system, ecological, social, governance)
- Opportunities and recommendations for future research and practice

This step generated a structured understanding of the key directions and limitations shaping the global NbS knowledge base.

Systematic Mapping

Task 1.3. systematically mapped empirical, place-based studies on urban NbS using searches in Scopus and Web of Science. Search strings combined variants of “nature-based solution(s)” with urban terms (city, cities, urban, town, metropolitan, peri-urban) in titles, abstracts, or author keywords. Results were limited to English-language journal articles published up until the end of 2023. After merging databases and removing duplicates and records without abstracts, 1,436 papers remained for screening; 1,031 papers met the criteria and were included in the final dataset.

Exclusion criteria—papers were included unless they:

- did not contain “nature-based solution(s)” in title, abstract, or keywords
- were not urban in focus
- were not place-based (e.g. conceptual only, no specific study location)

- contained more than ten case studies across multiple world regions (most of these papers were assessed in Task 1.2)

Each paper was screened on title and abstract by two reviewers ("yes", "no", or "maybe"), with conflicts resolved through discussion and, for "maybe" cases, full-text checks.

For the retained 1,031 papers, analyses combined manual and automated methods:

- Manual geographic tagging: Using information provided in titles, abstracts, and full texts, the team manually assigned each paper to a region, country, city/town, and district/neighbourhood where reported. Papers could receive multiple geographic tags if they covered more than one location.

- Bibliometric analysis: Using the bibliometrix package in R, we examined:
  - publication trends over time by region
  - the most frequent journals publishing on urban NbS
  - the most cited papers per region, and
  - the relationship between author affiliation regions and study locations (e.g. proportion of in-country vs external researchers).

These indicators were used to characterise patterns and asymmetries in global research production on urban NbS.

- Co-occurrence mapping: Vosviewer was used to generate keyword co-occurrence maps based on author-provided keywords. Papers were split into regional subsets according to the manual tagging, and the minimum keyword occurrence threshold was adjusted to reflect the number of papers in each region. The maps were used to identify dominant thematic clusters and co-occurring concepts across and within regions.

- Text mining: Text mining in R was guided by a predefined keyword codebook developed iteratively using expert input, case study data (Task 2), and reference frameworks (e.g. IUCN, NetworkNature). The codebook classified keywords under two main categories—ecosystem types (green, blue, brown, etc.) and societal challenges (climate, ecosystem health, wellbeing, inclusion, etc.)—and was refined through multiple validation rounds until achieving an average coding accuracy of 94.9%.

The resulting dataset provided the empirical foundation for visualizing geographic and thematic research patterns, highlighting where and how NbS are studied across regions.

Task 2 | Case study selection

The regional review of NbS practices aimed to develop a regional profile of urban NbS policy, planning and implementation in Europe through a set of representative case studies selected via an expert-based process. The selection followed four main steps (Figure 11).

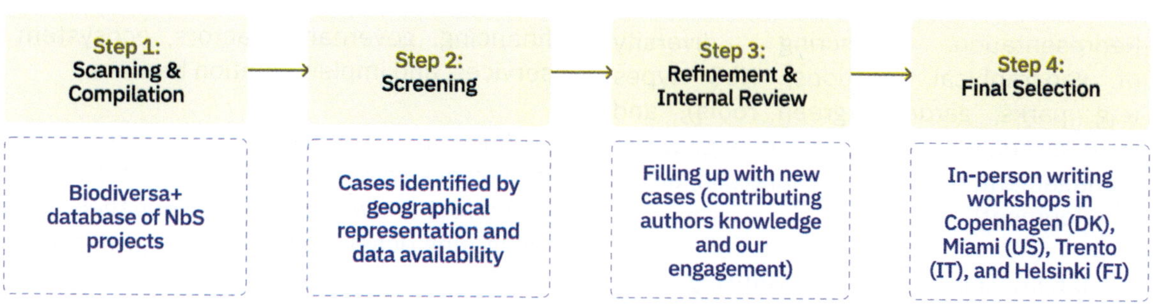

Figure 11: Four Main Steps in the Case Study Selection Process.

Step 1: Scanning and compilation

The initial pool of case studies was drawn from BiodivERsA-funded projects (2015–2021), prioritized for their robust data and accessibility. To assess representativeness, these cases were cross-checked against the NATURVATION project's Urban Nature Atlas, which catalogues NbS in 100 European cities (Figure 12). Additional cases were identified through grey and academic literature and through the networks of the NATURA regional team for Europe and its contributing authors.

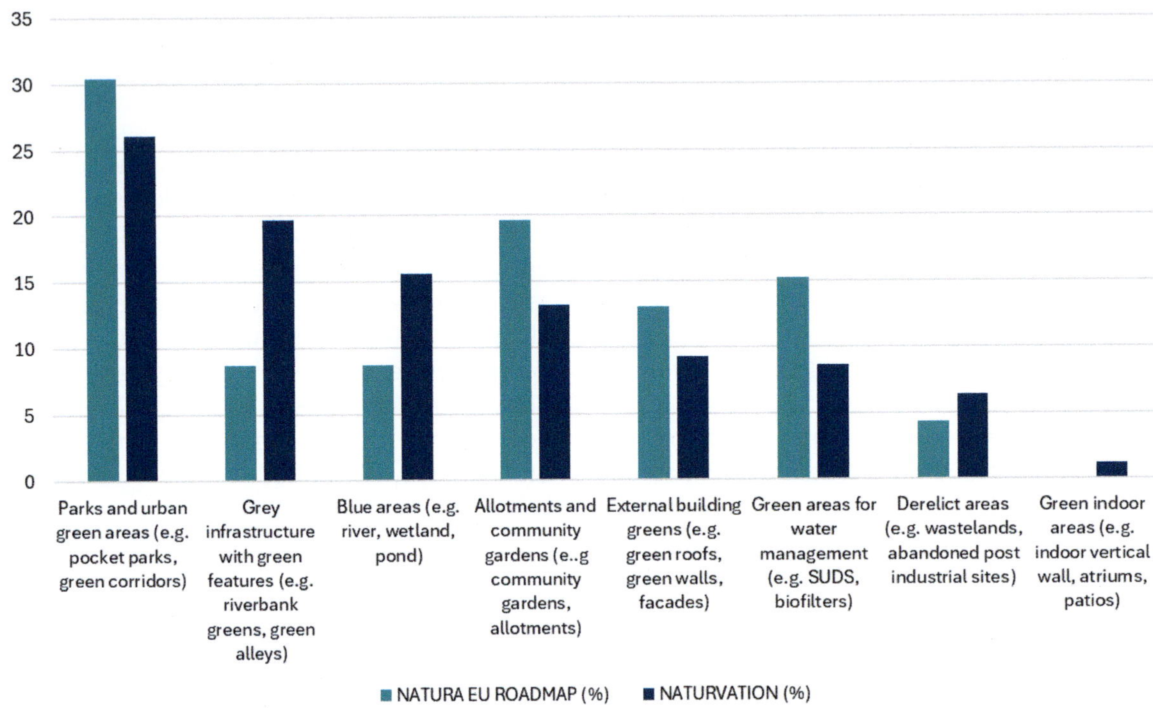

Figure 12: Comparison of the Preliminary Case Study Selection with NATURVATION's Urban Nature Atlas by Ecosystem Type, Highlighting Gaps and Potential Biases in Representation.

Step 2: Screening

A set of agreed criteria guided the screening process:

- Representation: ensuring diversity of geographical locations, NbS types (e.g. parks, gardens, green roofs), and societal challenges addressed (e.g. water management, regeneration, land use, urban development).
- Availability of information: prioritizing cases with sufficient and verifiable documentation to support detailed analysis of drivers, challenges, NbS types, intervention scales, financing, governance, actors, ecosystem services, and implementation barriers.
- Collaboration and prior knowledge: giving preference to cases where Node members had direct knowledge or prior engagement, facilitating richer contextual interpretation.

Step 3: Refinement and internal review

The preliminary pool was refined through iterative internal discussions to balance representativeness and thematic coverage. Identified geographic or sectoral gaps were addressed by sourcing additional case studies through extended networks. This process ensured broader coverage of Europe's varied socio-ecological and governance contexts.

Step 4: Final selection and synthesis

The final set of 22 case studies was validated through collaborative synthesis workshops held in Copenhagen (Denmark), Miami (USA), Trento (Italy), and Helsinki (Finland). These workshops gathered contributing authors to discuss findings and integrate insights into a coherent regional narrative for the roadmap.

The selected cases represent a geographically diverse sample spanning northern, western, central, and southern Europe Fewer examples were available from eastern, southeastern, and Baltic Europe, reflecting persistent language and data accessibility barriers. Addressing these gaps will be crucial for future efforts to build a more comprehensive picture of NbS implementation across Europe's urban contexts.

B Case Study Sites and NbS Typologies

Table 1: Europe case study sites and NbS typologies

Site	**Parks and urban green areas** (e.g. pocket parks, green corridors)	**Grey infra-structure with green features**	**Blue areas** (e.g. river wetland, pond, coastal ecosystems)	**Allotments and commu-nity gardens** (e.g. commu-nity gardens, allotments)	**External building greens** (e.g. green roofs, green walls, facades)	**Green areas for water manage-ment** (e.g. SUDS, biofilters)	**Derelict areas** (e.g. waste-lands, aban-doned postindus-trial sites)	**Green indoor areas** (e.g. indoor vertical wall, atriums, patios)
Viva Plan/ SGC	Type III					Type III		
SPARK	Type III		Type III	Type III		Type III		
CO-CO-NAT 2025	Type III	Type II	Type II			Type III		
Grønne Veje						Type III		
UNaLab (Tampere)	Type III		Type III		Type III	Type III		
UPSURGE (Belfast)	Type III			Type III				
UNaLab (Eindhoven)	Type III		Type III		Type III	Type III		
ProGIreg	Type III			Type III	Type III			
Urban GreenUp	Type III					Type III		
Clever Cities	Type III				Type III			
Coolschools	Type III			Type III				
CLEARING HOUSE							Type I	
BiodiverCity	Type III	Type III		Type III	Type III	Type III		
Connecting Nature (Glasgow)				Type II / Type III				
Connecting Nature (A Coruna)				Type II / Type III				
URBiNAT	Type III							
Krakov	Type III				Type III	Type III		
Municipal horticultur-al parks of Lisbon				Type III				
Rumon (Finale Emilia)				Type II				
Gâvres			Type II / Type III					
Südpanke	Type III		Type II / Type III					

C Summary of Europe Case Studies

Table 2: Summary of Europe case studies (cases EU_01 to EU_05)

ID (EU_)	Project name	Country	Societal problems addressed	Co-benefits	Impact scale	Urban typology	Post-intervention ecosystem category	Post-intervention ecosystem subcategory	NbS typology	NbS techniques I
01	The Hørgården Infrastructure Plan: Co-Designing Green Space Around Relational Values	Denmark	Flooding; Social inclusion	Increased biodiversity and their functions; Place regeneration & urban revitalization; stronger human-nature connections	Neigh-bour-hood/ commu-nity	Urban core	Green; Blue-green	Stormwater infrastructure; Green spaces (gen.)	Constructed & novel ecosystems	Recreational & private green space management; Urban green infrastructure construction; Stormwater absorption, retention & treatment
02	Healthy Park Activities, Rehabilitation, and Climate (SPARK): Park Revitalization in Aarhus	Denmark	Flooding	Increased biodiversity and their functions; Recreational benefits	Site	Urban core	Green	Green spaces (gen.); Meadows & lawns; Stormwater infrastructure	Restoration; Constructed & novel ecosystems	Recreational & private green space management; Forestation & urban tree strategies; Revegetation & planting (gen.); Stormwater absorption, retention & treatment
03	Connecta Congost Natura 2025: Rehabilitating the Congost River	Spain	Physical and mental health; Flooding	Increased biodiversity and their functions; Place regeneration & urban revitalization	City	Urban core	Green; Blue	Green spaces (gen.); Stormwater infrastructure; Rivers & waterways; Forests & trees	Restoration	Stormwater absorption, retention & treatment; Freshwater ecosystem creation, restoration & management
04	Grønne Veje: Stormwater Management in Copenhagen	Denmark	Flooding	Recreational benefits	Greater urban region	Peri-urban	Blue-green; Green	Green spaces (gen.); Stormwater infrastructure;	Constructed & novel ecosystems	Urban green infrastructure construction; Stormwater absorption, retention & treatment; Revegetation & planting (gen.)
05	UNaLab Tampere: Living Labs for Nature-Based Solutions	Finland	Flooding; Biodiversity loss; Water quality	Recreational benefits; Stronger human-nature connections	Neigh-bour-hood/ commu-nity	Urban core	Green; Blue-green	Green spaces (gen.); Roofs & walls; Stormwater infrastructure; Forests & trees	Restoration; Constructed & novel ecosystems	Recreational & private green space management; Wastewater treatment; Stormwater absorption, retention & treatment

ID (EU_)	Project name	Country	NbS technique II	Primary taxonomy	Domestic government role	Foreign government role	Non-governmental actors role	Private sector role	Academia role
01	**The Hørgården Infrastructure Plan: Co-Designing Green Space Around Relational Values**	Denmark	Stormwater management (gen.); Green park restoration	Nature-based Solutions	Primary funder; Project lead	No role	No role	Implementation partner	Research media and dissemination
02	**Healthy Park Activities, Rehabilitation, and Climate (SPARK): Park Revitalization in Aarhus**	Denmark	Trenches & ditches; Recreational green space management (gen.); Tree planting (outside of forests); Vegetation planting; Bioretention & infiltration basins	Nature-based Solutions	Primary funder	No role	No role	Support funder; Maintenance and/or stewardship; Implementation partner	No role
03	**Connecta Congost Natura 2025: Rehabilitating the Congost River**	Spain	Permeable pavement; Green streets; River restoration (gen.)	Ecological or natural infrastructure	Project lead	Primary funder	No role	Research media and dissemination	No role
04	**Grønne Veje: Stormwater Management in Copenhagen**	Denmark	Bioswales, biofiltration & rain gardens	Green infrastructure (GI)	Primary funder; Implementation partner	No role	No role	Implementation partner	No role
05	**UNaLab Tampere: Living Labs for Nature-Based Solutions**	Finland	Retention ponds & basins; Community gardens (non-agricultural); Nature trail construction & maintenance; Biofilters; Recreational green space management (gen.); Aquaponics; Microalgae systems (for biogas)	Nature-based Solutions	Project lead	Primary funder	No role	Implementation partner	No role

ID (EU_)	Project name	Country	Community role	Indigenous groups	Institutional instruments	Project timeline			Status	Budget	Implementation barriers	Implementation enablers
						Start	End	Duration (yrs)				
01	**The Hørgården Infrastructure Plan: Co-Designing Green Space Around Relational Values**	Denmark	Participant	0	Law; Policy; Plan (e.g., Master plan)	2019	2025	6	Done	5,000,000—10,000,000	Equity, inclusion & engagement	Equity-centered; Stakeholder engagement; Co-production approaches in management; Innovative and adapted design; Technical expertise
02	**Healthy Park Activities, Rehabilitation, and Climate (SPARK): Park Revitalization in Aarhus**	Denmark	Implementation partner	0	Plan (e.g., Master plan); International agreements	2014	2022	8	Done	5,000,000—10,000,000	Values and/or priorities conflict; External disruptions; Design adaptation and skill gaps	Good coordination or collaboration; Funding available
03	**Connecta Congost Natura 2025: Rehabilitating the Congost River**	Spain	Participant	0	Plan (e.g., Master plan)	2022	2025	3	Ongoing	1,000,000—5,000,000	Equity, inclusion & engagement	Equity-centered; Stakeholder engagement; Co-production approaches in management; Funding available; Innovative and adapted design; Technical expertise
04	**Grønne Veje: Stormwater Management in Copenhagen**	Denmark	Implementation partner	0	Plan (e.g., Master plan)	2018	2038	20	Ongoing	100,000,000—1,000,000,000	Design adaptation and skill gaps; Poor coordination or cross-sectoral integration	Equity-centered; Stakeholder engagement; Co-production approaches in management; Enabling policies and regulations
05	**UNaLab Tampere: Living Labs for Nature-Based Solutions**	Finland	Participant	0	Program	2017	2022	5	Done	1,000,000—5,000,000	Design adaptation and skill gaps	Equity-centered; Stakeholder engagement; Co-production approaches in management; Enabling policies and regulations

Table 3: Summary of Europe case studies (cases EU_06 to EU_10)

ID (EU_)	Project name	Country	Societal problems addressed	Co-benefits	Impact scale	Urban typology	Post-intervention ecosystem category	Post-intervention ecosystem subcategory	NbS typology	NbS techniques I
06	**UPSURGE: Soil Remediation and Urban Agriculture in Belfast**	Northern Ireland, UK	Social inclusion; Land degradation	Soil stability; Fertility and quality; Recreational benefits; Place regeneration & urban revitalization	Site	Urban core	Green	Agricultural & food spaces	Restoration	Recreational & private green space management; Urban agricultural practices; Erosion risk reduction & soil remediation
07	**UNaLab Eindhoven: Living Labs for Nature-Based Solutions**	The Netherlands	Heat; Flooding	Urban cooling; Increased biodiversity and their functions	City	Urban core	Green; Blue	Green spaces (gen.); Roofs & walls; Corridors; Rivers & waterways	Restoration; Constructed & novel ecosystems	Forestation & urban tree strategies; Urban green infrastructure construction; Revegetation & planting (gen.)
08	**ProGIreg Living Lab for Green Infrastructure and Post-Industrial Regeneration**	Croatia	Nature connection; Physical and mental health; Heat	Recreational benefits; Flood mitigation	Site	Urban core	Green	Agriculture & food spaces; Green spaces (gen.); Roofs & walls; Corridors; Forests & trees	Restoration; Constructed & novel ecosystems	Recreation & private green space management; Urban agricultural practices; Urban green infrastructure construction; Wastewater treatment; Forestation & urban tree strategies
09	**UrbanGreenUp: Green Infrastructure in Liverpool**	United Kingdom	Heat; Flooding; Water quality; Nature connection	Increased biodiversity and their functions; Carbon sequestration; Recreational benefits; Green jobs and economic benefits	City	Urban core	Green; blue-green	Green spaces (gen.); Corridors; Forests & trees; Stormwater infrastructure; Roofs & walls; Productive blue spaces; Agriculture & food spaces	Restoration; Constructed & novel ecosystems	Urban green infrastructure construction; Forestation & urban tree strategies; Wastewater treatment; Revegetation & planting (general); Stormwater absorption, retention & treatment
10	**CLEVER Cities: Urban Greening and Co-design in Milan**	Italy	Heat; Biodiversity loss; Flooding	Recreational benefits	City	Urban core	Green	Roofs & walls; Green spaces (gen.);	Restoration; Constructed & novel ecosystems	Urban green infrastructure construction; Recreational & private green space management

ID (EU_)	Project name	Country	NbS technique II	Primary taxonomy	Domestic government role	Foreign government role	Non-governmental actors role	Private sector role	Academia role
06	**UPSURGE: Soil Remediation and Urban Agriculture in Belfast**	Northern Ireland, UK	Horticulture; Soil remediation; Recreational green space management (gen.); Vegetable gardening	Nature-based Solutions	Project lead	Primary funder	Implementation partner	Unknown	Research media and dissemination; Specialized knowledge
07	**UNaLab Eindhoven: Living Labs for Nature-Based Solutions**	The Netherlands	Tree planting (outside of forests); Vegetation planting; Green walls & façades; Green roofs	Nature-based Solutions	Project lead	Primary funder	No role	Implementation partner	Research media and dissemination
08	**ProGIreg Living Lab for Green Infrastructure and Post-Industrial Regeneration**	Croatia	Community gardens (non-agricultural); Aquaponics; Green roofs; Green walls & façade; Green alleys & corridors; Urban farming (gen.); Water treatment (gen.); Tree planting (outside of forests)	Green infrastructure (GI)	Project lead	Primary funder	Participant; Research media and dissemination	Implementation partner	Research media and dissemination
09	**UrbanGreenUp: Green Infrastructure in Liverpool**	United Kingdom	Green alleys & corridors; Tree planting (outside of forests); Bioswales, biofiltration & rain gardens; Floating gardens & islands; Bird, insect & mammal shelters; Soil remediation; Vegetation planting; Mangrove conservation	Nature-based Solutions	Project lead	Primary funder	Unknown	Participant	Unknown
10	**CLEVER Cities: Urban Greening and Co-design in Milan**	Italy	Green roofs; Green walls & façades; Recreational green space management (gen.); Green park restoration; Native & indigenous species planting	Nature-based Solutions	Project lead; Primary funder	Support funder	Implementation partner	Implementation partner	Research media and dissemination

ID (EU_)	Project name	Country	Community role	Indigenous groups	Institutional instruments	Project timeline			Status	Budget	Implementation barriers	Implementation enablers
						Start	End	Duration (yrs)				
06	**UPSURGE: Soil Remediation and Urban Agriculture in Belfast**	Northern Ireland, UK	Participant	0	Plan (e.g., Master plan)	2021	2025	4	Done	500,000—1,000,000	Unknown	Equity-centered; Stakeholder engagement; Co-production approaches in management; Enabling policies and regulations; Funding available; Education, awareness, and advocacy
07	**UNaLab Eindhoven: Living Labs for Nature-Based Solutions**	The Netherlands	Participant	0	Program	2017	2022	5	Done	1,000,000—5,000,000	Equity, inclusion & engagement; Poor coordination or cross-sectoral integration	Funding available; Innovative and adapted design; Technical expertise
08	**ProGIreg Living Lab for Green Infrastructure and Post-Industrial Regeneration**	Croatia	Participant	0	Policy	2018	2021	3	Done	1,000,000—5,000,000	Implementation delays; External disruptions	Enabling policies and regulations; Equity-centered; Stakeholder engagement; Co-production approaches in management;
09	**UrbanGreenUp: Green Infrastructure in Liverpool**	United Kingdom	Participant	0	Unknown	2017	2022	5	Done	10,000,000—50,000,000	Monitoring, evaluation and learning gaps	Equity-centered; Stakeholder engagement; Co-production approaches in management; Good coordination or collaboration; Funding available; Innovative and adapted design; Technical expertise
10	**CLEVER Cities: Urban Greening and Co-design in Milan**	Italy	Participant	0	Plan (e.g., Master plan)	2018	2023	5	Done	10,000,000—50,000,000	Poor coordination or cross-sectoral integration; Poor policy & regulatory frameworks; Values and/or priorities conflicts	Funding available; Innovative and adapted design; Technical expertise

Table 4: Summary of Europe case studies (cases EU_11 to EU_15)

ID (EU_)	Project name	Country	Societal problems addressed	Co-benefits	Impact scale	Urban typology	Post-intervention ecosystem category	Post-intervention ecosystem subcategory	NbS typology	NbS techniques I
11	**Opération-Récréation: Schoolyard Greening in Brussels**	Belgium	Flooding; Biodiversity loss; Social well-being; Nature connection	Place regeneration & urban revitalization	Site	Urban core	Green	Green spaces (gen.); Agriculture & food spaces	Restoration	Recreational & private green space management; Forestation & urban tree strategies; Species conservation & monitoring; Revegetation & planting (gen.); Urban agricultural practices
12	**ClearingHouse: Post-Industrial Urban Forests in Gelsenkirchen**	Germany	Nature connection; Physical and mental health	Place regeneration & urban revitalization; Community engagement & social cohesion; Reinforcement of socio-ecological knowledge	Greater urban region	Peri-urban	Green	Forests & trees	Restoration	Terrestrial rehabilitation & site remediation
13	**The BiodiverCity Project: Biodiversity Conservation in Malmö**	Sweden	Biodiversity loss; Social inclusion	Flood mitigation; Human health and well-being; Urban cooling; Reduced infrastructure cost; Air quality	City	Urban core	Green; Blue-green	Green spaces (gen.); Stormwater infrastructure; Roofs & walls; Riparian areas	Restoration; Constructed & novel ecosystems	Urban green infrastructure construction; Urban agricultural practices; Stormwater absorption, retention & treatment
14	**Connecting Nature: An Urban Gardens Green Network**	Spain	Social well-being; Nature connection	Community engagement & social cohesion; Reinforcement of socio-ecological knowledge	City	Urban core	Green	Agriculture & food spaces	Constructed & novel ecosystems	Urban agricultural practices
15	**URBiNAT: Sofia's "Healthy Corridor"**	Bulgaria	Nature connection; Physical and mental health	Place regeneration & urban revitalization;	Neigh-bour-hood/commu-nity	Urban core	Green	Corridors; Green spaces (gen.)	Restoration; Constructed & novel ecosystems	Urban green infrastructure construction; Revegetation & planting (gen.);

ID (EU_)	Project name	Country	NbS technique II	Primary taxonomy	Domestic government role	Foreign government role	Non-govern-mental actors role	Private sector role	Academia role
11	**Opération-Récréation: Schoolyard Greening in Brussels**	Belgium	Green park restoration; Tree planting (outside of forests); Bird, insect & mammal shelters; Vegetation planting; Vegetable gardening	Nature-based Solutions	Project funder; Implementa-tion partner	No role	No role	Unknown	No role
12	**ClearingHouse: Post-Industrial Urban Forests in Gelsenkirchen**	Germany	Post-industrial and contaminated site rehabilitation	Nature-based Solutions	Project lead	Primary funder	Unknown	Unknown	Research media and dissemination
13	**The BiodiverCity Project: Biodiversity Conservation in Malmö**	Sweden	Green roofs; Green walls & façades; Green alleys & corridors; Community gardens (non-agricultural); Sustainable urban drainage systems (SUDS); Afforestation	Nature-based Solutions	Project lead; Support funder	Primary funder	Support funder	Implementa-tion partner	Research media and dissemination
14	**Connecting Nature: An Urban Gardens Green Network**	Spain	Urban farming (gen.)	Nature-based Solutions	Project lead; Primary funder	Support funder	Implementation partner	Implementa-tion partner	Unknown
15	**URBiNAT: Sofia's "Healthy Corridor"**	Bulgaria	Constructed linear park; Recreational green space management (gen.); Green alleys & corridors;	Nature-based Solutions	Project lead; Support funder	Primary funder	Unknown	Unknown	Research media and dissemination

ID (EU_)	Project name	Country	Community role	Indigenous groups	Institutional instruments	Project timeline			Status	Budget	Implementation barriers	Implementation enablers
						Start	End	Duration (yrs)				
11	**Opération-Récréation: Schoolyard Greening in Brussels**	Belgium	Participant	0	Plan (e.g., Master plan)	2021	2024	3	Done	5,000,000–10,000,000	High costs & economic risks; Implementation delays	Good coordination or collaboration; Equity-centered; Stakeholder engagement; Co-production approaches in management;
12	**ClearingHouse: Post-Industrial Urban Forests in Gelsenkirchen**	Germany	Participant	0	Plan (e.g., Master plan)	2019	2024	5	Done	500,000–1,000,000	Values and/or priorities conflicts; Funding gaps	Enabling policies and regulations;
13	**The BiodiverCity Project: Biodiversity Conservation in Malmö**	Sweden	Participant	0	Policy	2012	2018	6	Done	1,000,000–5,000,000	Design adaptation and skill gaps	Enabling policies and regulations; Funding available; Education, awareness, and advocacy
14	**Connecting Nature: An Urban Gardens Green Network**	Spain	Implementation partner	0	Policy	2018	2018	0	Done	10,000,000–50,000,000	Resistance to change & cultural barriers; Values and/or priorities conflicts; Poor coordination or cross-sectoral integration	Enabling policies and regulations; Funding available; Environmental stewardship
15	**URBiNAT: Sofia's "Healthy Corridor"**	Bulgaria	Participant	0	Urban strategy	2018	2024	6	Done	10,000,000–50,000,000	Unknown	Equity-centered; Stakeholder engagement; Co-production approaches in management; Good coordination or collaboration; Innovative and adapted design; Technical expertise

Table 5: Summary of Europe case studies (cases EU_16 to EU_21)

ID (EU_)	Project name	Country	Societal problems addressed	Co-benefits	Impact scale	Urban typology	Post-intervention ecosystem category	Post-intervention ecosystem subcategory	NbS typology	NbS techniques I
16	**Flower Power: A Community-run Wildflower Nursery in Glasgow**	United Kingdom	Social inclusion; Biodiversity loss; Heat	Place regeneration & urban revitalization;	Neigh-bour-hood/ Commu-nity	Urban core	Green	Green spaces (gen.)	Restoration	Recreational & private green space management; Species conservation & monitoring; Wetland conservation, restoration & construction
17	**Green Krupnicza: Blue-Green Infrastructure in Kraków**	Poland	Heat	Flood mitigation	Street	Urban core	Green; Blue-green	Green spaces (gen.); Stormwater infrastructure; Forests & trees; Marine (gen.)	Constructed & novel ecosystems	Stormwater absorption, retention & treatment; Forestation & urban tree strategies; Mangrove conservation & restoration
18	**Lisbon's Strategy for Urban Agriculture**	Portugal	Food security; Social inclusion	Water quality; Recreational benefits	Neigh-bour-hood/ Commu-nity	Urban core	Green	Agriculture & food spaces; Green spaces (gen.);	Restoration	Urban agricultural practices; Recreational & private green space management; Wetland conservation, restoration & construction
19	**Finale Emilia Food Forest Supported by the Rumon Association**	Italy	Biodiversity loss; Social inclusion	Green jobs and economic benefits	Site	Peri-urban	Green	Agriculture & food spaces; Forests & trees	Restoration	Urban agricultural practices; Forestation & urban tree strategies; Revegetation & planting (gen.)
20	**Intercommunal Dune Management in Brittany**	France	Land degradation; Flooding	Increased biodiversity and their functions	Greater urban area	Outside the city with urban impacts	Blue	Coastal (gen.)	Restoration	Erosion risk reduction & soil remediation; Revegetation & planting (gen.)
21	**Green corridor alongside the Südpanke stream**	Germany	Physical and mental health; Nature connection	Place regeneration & urban revitalization; Recreational benefits; Human health and well-being; Increased biodiversity and their functions	Site	Urban core	Blue; Green	Rivers & waterways; Forests & trees;	Restoration	Urban green infrastructure construction; Frashwater ecosystem creation, restoration & management

ID (EU_)	Project name	Country	NbS technique II	Primary taxonomy	Domestic government role	Foreign government role	Non-governmental actors role	Private sector role	Academia role
16	**Flower Power: A Community-run Wildflower Nursery in Glasgow**	United Kingdom	Recreational green space management (gen.); Native & indigenous species planting; Community gardens (non-agricultural); Floating gardens & islands	Nature-based Solutions	Project lead; Support funder	Primary funder	Research media and dissemination	Unknown	Research media and dissemination
17	**Green Krupnicza: Blue-Green Infrastructure in Kraków**	Poland	Bioswales, biofiltration & rain gardens; Vegetation planting	Nature-based Solutions	Primary funder; Project lead	No role	Implementation partner	Support funder; Maintenance and/or stewardship	Research media and dissemination
18	**Lisbon's Strategy for Urban Agriculture**	Portugal	Horticulture; Urban farming (gen.); Recreational green space management (gen.); French filter drains	Nature-based Solutions	Project lead; Primary funder	No role	No role	Unknown	No role
19	**Finale Emilia Food Forest Supported by the Rumon Association**	Italy	Food forests; Tree planting (outside of forests); Vegetation planting; Vegetable gardening	Food forest	Support funder	No role	Project lead	Primary funder	No role
20	**Intercommunal Dune Management in Brittany**	France	Dune restoration; Stabilization; Vegetation planting; Erosion risk reduction	Nature-based Solutions	Primary funder; Project lead	No role	No role	No role	Specialized knowledge
21	**Green corridor alongside the Südpanke stream**	Germany	Daylighting; Recreational green space management (gen.)	Green infrastructure (GI)	Implementation partner; Primary funder	No role	Implementation partner	Implementation partner	No role

ID (EU_)	Project name	Country	Community role	Indigenous groups	Institutional instruments	Project timeline			Status	Budget	Implementation barriers	Implementation enablers
						Start	End	Duration (yrs)				
16	**Flower Power: A Community-run Wildflower Nursery in Glasgow**	United Kingdom	Maintenance and/or stewardship	0	Plan (e.g., Master plan)	2017	2022	5	Done	10,000,000–50,000,000	Funding gaps	Good coordination or collaboration; Education, awareness, and advocacy
17	**Green Krupnicza: Blue-Green Infrastructure in Kraków**	Poland	Implementation partner; Maintenance and/or stewardship	0	Urban strategy	2022	2023	1	Done	1,000,000–5,000,000	Unknown	Equity-centered; Stakeholder engagement; Co-production approaches in management; Good coordination or collaboration; Funding available
18	**Lisbon's Strategy for Urban Agriculture**	Portugal	Implementation partner	0	Urban strategy	2011		14	Ongoing	Unknown	Equity, inclusion & engagement	Regular maintenance; Education, awareness, and advocacy
19	**Finale Emilia Food Forest Supported by the Rumon Association**	Italy	Project lead; Primary funder; Maintenance and/or stewardship	0	Unknown	2022		3	Ongoing	< 100,000	Dependency on donors	Cultural integration; Institutional capacity (e.g., decentralized approaches); Education, awareness, and advocacy; Environmental stewardship
20	**Intercommunal Dune Management in Brittany**	France	Specialized knowledge	0	Law	2018		7	Ongoing	500,000–1,000,000	Poor coordination or cross-sectoral integration; Resistance to change & cultural barriers	Enabling policies and regulations
21	**Green corridor alongside the Südpanke stream**	Germany	No role	0	Policy	2020	2021	1	Done	10,000,000–50,000,000	Poor ecological health; Values and/or priorities conflicts	Enabling policies and regulations; Funding available; Innovative and adapted design; Technical expertise

D Case Study Write-ups

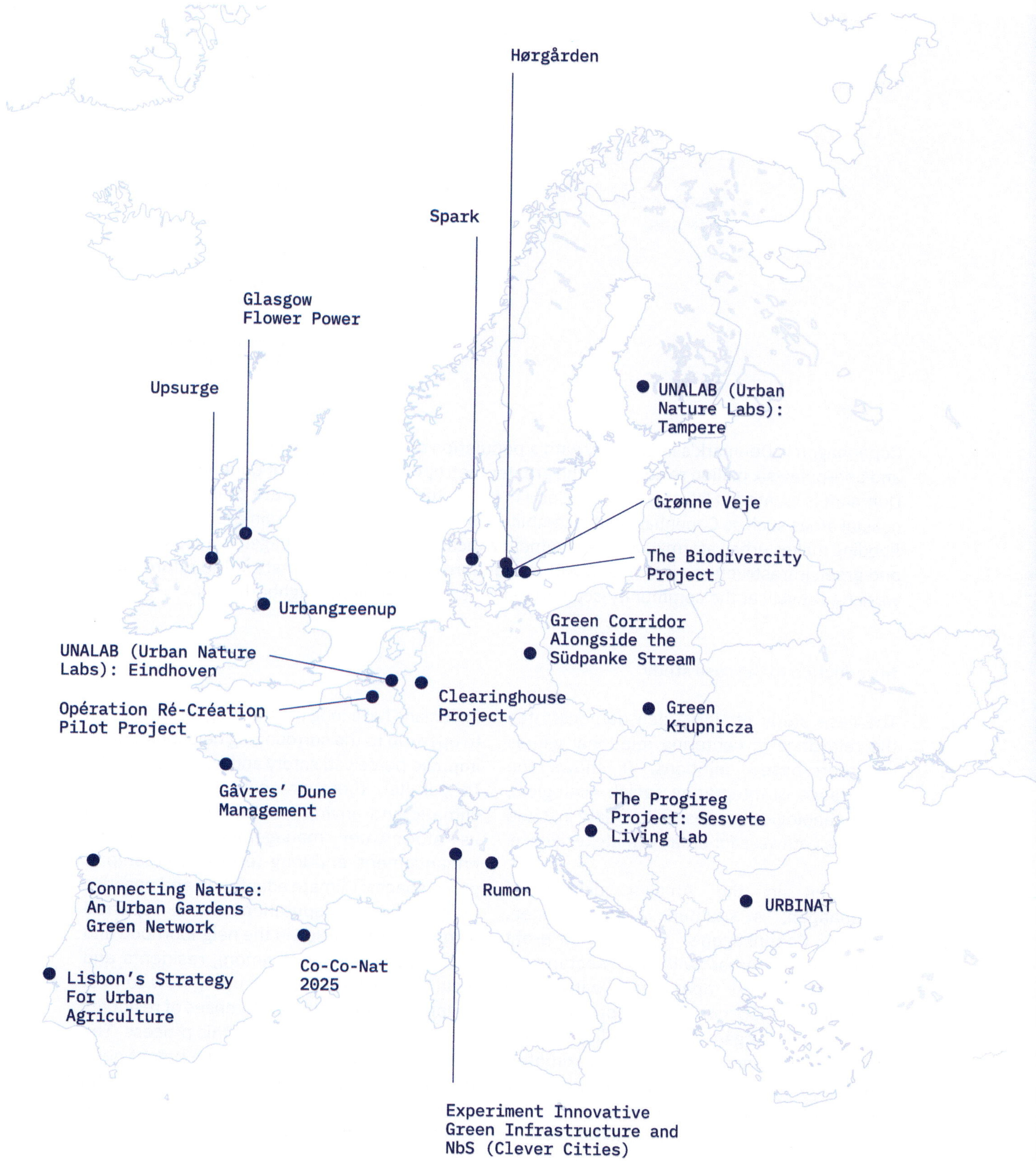

EU_01

The Hørgården Infrastructure Plan: Co-Designing Green Space Around Relational Values

Context Overview

LOCATION

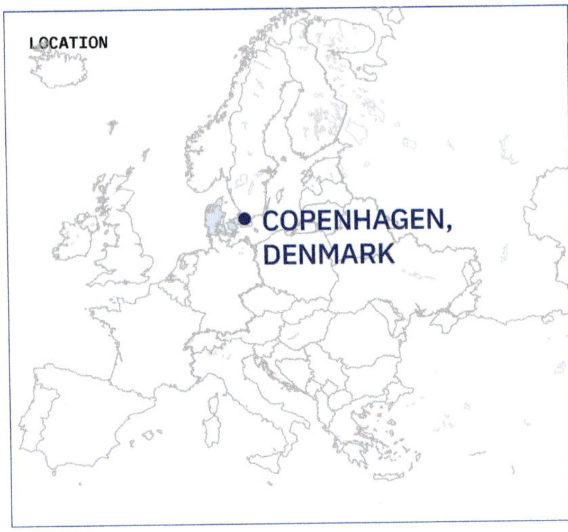

NBS TYPOLOGY Constructed & Novel Ecosystems	ECOSYSTEM TYPE Stormwater infrastructure; Green spaces (gen.)
IMPACT SCALE Neighborhood/Community	URBAN TYPOLOGY Urban core
SOCIAL CHALLENGES ADDRESSED Flooding; Social inclusion	CO-BENEFITS Increased biodiversity and their functions; Place regeneration & urban revitalization; Stronger human-nature connections

Copenhagen is Denmark's largest city with a population of 650,000 in the municipal boundaries and one-point-six million residents in the greater metropolitan area (City of Copenhagen, 2023). Denmark is located in Northern Europe, and is characterized by a moderate coastal climate. In coastal areas such as Copenhagen, the combination of increasingly heavy rain events and coastal flooding makes urban stormwater management a focal challenge. Copenhagen is increasing blue and green infrastructure city-wide to withstand more frequent cloudbursts in the future and is working actively at the community-scale to realize climate resilient neighborhoods.

Introduction to the case study

The case study of Hørgården demonstrates the relevance of centering relational values in nature-based solutions. It shows the importance of integrating social, ecological, and technological processes to create just and inclusive pathways for urban climate resilience.

Hørgården in the Amager district of Copenhagen, is situated in Urbanplanen, one of Copenhagen's largest non-profit social housing areas with 450 hectares of mostly multi-story apartments with 6000 residents and 50 hectares of public urban green space. Hørgården is a smaller housing block within Urbanplanen with approximately 1600 residents, many of whom are struggling socio-economically due to lesser education or integration into the Danish labor market. Politicians have long been calling for Hørgården to open-up to the surrounding neighborhood to improve perceived safety and social resilience. In parallel, there are calls for addressing climatic and environmental challenges such as stormwater management, biodiversity enhancement, and long-term stewardship of green spaces. Climate adaptation and negative neighborhood reputation have spurred significant renovation in the neighborhood and have raised questions among residents and local decision-makers regarding how to best engage the voices and preferences of residents in nature-based solutions in this process. The case of Hørgården addresses the urgent need for climate adaptation and environmental justice with a community-based approach to planning for nature-based solutions.

Nature-based Solutions characteristics

The Hørgården Infrastructure Plan, funded with 67 million DKK (10 million USD) from Landsbyggefonden in 2019, aims to improve the neighborhood's reputation and its physical and ecological infrastructure through community-driven projects. Local residents, housing administrators, and a residential council were instrumental in securing the funding. They have worked together to enhance connections within the neighborhood and create community-shaping areas, allowing residents to co-design green spaces that reflect their values and preferences while the green common areas are renovated due to a stormwater management and sewer upgrade project. Community engagement programs have successfully identified residents' desires for these areas' functions, uses, and names. The stormwater management project is mandated by the City of Copenhagen in line with its Cloudburst Plan (2012) to increase the service level of the sewer system and coordinated by the utilities company HOFOR. This project involves replacing the current sewer system and separating the on-site stormwater from the sewer to instead send the stormwater to an underground cloudburst tunnel.

Governance

The coordinated effort to plan and implement intersecting nature-based solutions projects for urban and ecological sustainability in Hørgården is facilitated by a multi-layered green governance structure. This structure consists of four layers:

1. Local Resident Groups: Hørgården has active resident groups participating in both formal and informal green initiatives;
2. Amagerplanen Program: This social masterplan supports social and ecological development through partnerships between social workers and local grassroots initiatives, with success measured against national policies;
3. City of Copenhagen: The city aims to enhance urban nature for green open spaces and climate adaptation while also prioritizing working opportunities for unemployed and under-skilled residents, particularly in marginalized neighborhoods like Hørgården;
4. Danish State "Ghetto Plan": While the diverse on-site actors are successfully and carefully engaging and working with local residents to realize policy demands that support that residents' relationships to place and community, the city-wide climate adaptation strategies may have unintended consequences for low-income renter families, especially if costly measures, such as the separation of stormwater from sewage, are mandated for not-for-profit building owners such as Hørgården.

The municipality is aware of this increase in housing prices but has not stopped it because their measure for success in the climate adaptation program in Copenhagen is sewer service level, which does not relate to the socio-economic resilience of residents. The coordination of these projects points to the conflicts and opportunities in implementing nature-based solutions, emphasizing the power of green governance that respects the community 's relations and values in the face of the pit-falls of techno-managerialism. Although all procedures have been followed in full compliance with regulations and legislation, and the mandate has been approved by policymakers, a missing link remains between housing tenure, a technocratic approach to climate adaptation, and its impact on marginalized communities. Policymakers should take this adaptation gap seriously as a key barrier to a just and equitable transition to climate resilience.

Outcome

The implementation of nature-based solutions in Hørgården and Urbanplanen is happening in phases. Phase One involves renovating Remiseparken, a central park in Urbanplanen. The renovation has been praised for enhancing climate adaptation, biodiversity, and social inclusion, and for making the park more accessible to Copenhagen. However, residents felt their perspectives were overlooked, as the redesign prioritized political goals of rebranding and safety improvements. Implementation in Urbanplanen, including Hørgården, focuses on better integrating residents' connections to their place and community into the design and engagement processes. The Hørgården Infrastructure Plan, along with stormwater management and urban densification projects, aims to address residents' concerns and visions during this period of change.

Two housing association staff members facilitate resident engagement, acting as liaisons between residents, the residents' association, researchers, architects, and the City of Copenhagen. They have successfully secured approval for "community-shaping areas," allowing residents to co-design green spaces reflecting their values and relationships. Over the past three years, facilitators have conducted community engagement programs to understand the diverse meanings, desired functions, and preferred names for these spaces. Additionally, researchers mapped residents' place values in Hørgården before new sewer systems and stormwater management projects began in 2023-2024.

Residents identified important social spots, biodiversity areas, and cultural heritage sites, highlighting areas critical for safety, recreation, and local culture. This mapping exercise ensured key sites were preserved and green spaces were reshaped to align with residents' values and cultural needs. The combined community engagement and research resulted in diverse green space designs supporting community relations, contributing to a resilient and sustainable urban and ecological plan. In Hørgården, nature-based solutions are addressed through place-based approaches that enhance community resilience and support residents' connections to their environment. Collaborating with the city and residents' board, designers and planners integrate local values into community areas while meeting regulatory requirements. Facilitators help align residents' values with these regulations, promoting sustainability.

This inclusive process strengthens relationships between residents and infrastructural changes by incorporating local narratives into redevelopment plans. Researchers, designers, and social workers engage residents to reshape their connections to nature and each other. This fosters trust and dialogue, crucial for long-term redevelopment. Construction of the new sewer system starts in January 2024, with the landscape design set for 2024 and 2025. Despite ongoing tensions, dialogue helps incorporate diverse values into the design, supporting resilient and just nature-based solutions.

Hørgården exemplifies the transformative governance advocated by policymakers and academics, highlighting that place-based approaches for community resilience require sensitivity to local perceptions, relational care, and significant time and resources to engage communities in planning processes. Additionally, the coordination of these projects points to the conflicts and opportunities in implementing nature-based solutions, emphasizing the power of green governance that respects the community's relations and values in the face of the pit-falls of techno-managerialism. Future focus should consider the missing link between housing tenure and technocratic management of climate adaptation.

Writing: Natalie Gulsrud
Review: Veronica Olivotto
Copy Editing: Veronica Olivotto
Design: Phuong Anh Bui

References:

1. City of Copenhagen (2023) Fremtidens klimavenlige hovedsta. Available at: https://www.kk.dk/sites/default/files/2024-05/Kommuneplanstrategi%20 2023_endelig%20version_feb2024-a.pdf.

2. Gulsrud, N. et al. (2023) 'Relational Values as a Nature-Based Solution in Copenhagen, Denmark', in A. Rubbo et al. (eds) Design for Resilient Communities. Cham: Springer International Publishing, pp. 319–331. Available at: https://doi.org/10.1007/978-3-031-36640-6_24.

3. Jørgensen, G. et al. (2022) 'Nature-based climate adaptation projects, their governance and transitional potential-cases from Copenhagen', Frontiers in Sustainable Cities, 4. Available at: https://doi.org/10.3389/frsc.2022.906960.

4. Manzo, L.C. et al. (2023) 'Using Senses of Place to Help Communities Navigate Place Disruption and Uncertainty', Landscape Journal, 42(1), pp. 37–52. Available at: https://doi.org/10.3368/lj.42.1.37.

5. Maurer, M. et al. (2023) 'A social-ecological-technological system approach to just nature-based solutions: A case of digital participatory mapping of meaningful places in a marginalized neighborhood in Copenhagen, Denmark', Urban Forestry & Urban Greening, 89, p. 128120. Available at: https://doi.org/10.1016/j.ufug.2023.128120.

6. Gulsrud, N., Garcia Antonez, O., Stahl-Olafsson, A., Bergen-Jensen, M. (Forthcoming/under revision) Housing Tenure, Climate Resilience, and Ecological Segregation. Proceedings of the National Academy of Science.

EU_02 Healthy Park Activities, Rehabilitation, and Climate (SPARK): Park Revitalization in Aarhus

Context Overview

LOCATION

NBS TYPOLOGY	ECOSYSTEM TYPE
Restoration; Constructed & novel ecosystems	Green spaces (gen.); Meadow & lawns; Stormwater infrastructure
IMPACT SCALE	URBAN TYPOLOGY
Site	Urban core
SOCIAL CHALLENGES ADDRESSED	CO-BENEFITS
Flooding	Increased biodiversity and their functions; Recreational benefits

Aarhus is the second largest city in Denmark, with a population of 290,598 inhabitants (Statistics Denmark 2023). Established in what used to be a brackish fjord, Aarhus is currently embedded in remnants of the former forest, which defines the northern and southern limits of the city. Located in the southern part of the city, close to one of these forest remnants, is Marselisborg Hospital, which served as the epidemiological hospital of the city from the late 19th century. Nowadays, Marselisborg Center is a national and international reference for rehabilitation, hosting more than 25 organizations related to human health (Handberg et al., 2018).

Introduction to the case study

SPARK is a park revitalization project that aims at increasing stormwater absorption as well as providing biodiversity and outdoor recreational opportunities to Marselisborg Center in Aarhus. Originally, the hospital had an enclosed hospital garden disconnected from surrounding streets by dense shrubbery and hedges. This created an inviting environment for the general public groups making illicit uses of the space. Moreover, in conjunction with the inadequate maintenance and care the park received, local residents perceived it as unwelcoming and felt insecure when walking through or next to it. Besides, Marselisborg Center is situated in an elevated area in the southern part of the city, which means that on-site stormwater management is key to protect surrounding, low-lying areas from flooding (Larsen, 2018).

SPARK stands for Healthy Park Activities, Rehabilitation, and Climate. The main objectives of this project (2014—2022) were fourfold; 1) provide outdoor rehabilitation facilities for patients in the Marselisborg Center; 2) offering recreational amenities to the more than 15,000 inhabitants residing at a walking distance from the site; 3) handle stormwater locally, and; 4) biodiversity conservation. Today, the park spans 72,000 square meters of urban green space, hosting 25 organizations specializing in various rehabilitation services (Handberg et al., 2018).

Nature-based Solutions characteristics

In order to handle 100-year cloudburst events, the topography of the former park and open space between the existing buildings were modified completely into a system of terrain elevations and depressions. This made it possible to create 24 differentiated areas hosting a variety of recreational as well as outdoor rehabilitation purposes (J. S. Johansen & C. Løvschall, personal communication 2023). Linking these areas is an interconnected network of vegetated infiltration basins, small depressions, and infiltration trenches that direct the water collected by the pipes of the different buildings into a small lake. In these activity areas, there are different materials and structures that the visitors and patients of the center can use for physical activity, outdoor rehabilitation and recreation, resting and socializing, outdoor gyms, rehabilitation structures as well as designated walking and running routes (A. Figueras, personal communication, 2024). Bushes, 1,300 trees, plant and vegetable beds, as well as meadows with annuals and perennials were implemented on the site to create new habitats for biodiversity.

Governance

SPARK is part of a wider urban development plan that began in 2014 and ended in 2022. The SPARK project was allocated a budget of approximately 55 million DKK (7.1 million EUR or 8.2 million USD). Key collaborators included Realdania, Mærsk, Marselisborg Center, Region Midtjylland, Aarhus Vand, and the Municipality of Aarhus. The Marselisborg Center is responsible for covering the ongoing maintenance and strategic management expenses of the park (J. S. Johansen & C. Løvschall, personal communication 2023). The project counted on the active engagement of more than 400 volunteering local residents throughout the different stages of the project (J. S. Johansen & C. Løvschall, personal communication 2023).

Outcome

At present, the park serves as a hub for diverse user groups engaging in various activities. Although occasional conflicts have arisen, such as concerns about smoke emanating from the fire pit, substantial issues between different user groups have not materialized. The overall perception of the project is overwhelmingly positive, as evidenced by the residents and partners alike, signifying a high level of success in its implementation. The project, however, faced some challenges at early planning stages. For example, the original vision aimed to connect the park with adjacent streets and the shared courtyards of the surrounding housing block (karré in Danish), creating a seamless integration with the urban landscape and encouraging local residents to use the space. However, this idea was ultimately abandoned due to opposition from resident associations, who prioritized maintaining privacy in their communal courtyards.

In addition, the initial project proposal had a stronger focus on rehabilitation and recreational functions. The funding bodies did not respond favorably to this approach, advising a shift towards emphasizing multifunctionality. Specifically, they recommended integrating additional environmental functions, such as stormwater management and biodiversity support. The new multifunctional aspect of the project favored its acceptance and indeed attracted more funding bodies to it. Furthermore, the involvement of local residents faced challenges due to the impact of COVID-19, resulting in a diminished role compared to the initial vision outlined by the project. Lastly, the significant topographic alterations necessary to integrate stormwater features mandated the removal of around 100 large trees. This action, arguably detrimental for local biological and cultural values, raises concerns about its sustainability and may be deemed as an unsustainable practice, pointing to a trade-off of urban nature-based solutions requiring heavy terrain modifications.

Community garden in Marselisborg Center

Green infrastructure at Marselisborg Center

Outdoor rehabilitation infrastructure in Marselisborg Center

Writing: Oriol Garcia-Antúnez
Review: Veronica Olivotto
Copy Editing: Veronica Olivotto
Design: Phuong Anh Bui

References:

1. A. Figueras (2024). AMPHI consult. Personal communication (January 5th, 2024).

2. Handberg, C., Mygind, O. and Johansen, J.S. (2019) 'Lessons learnt on the meaning of involvement and co-creation in developing community-based rehabilitation', Disability and Rehabilitation, 41(25), pp. 3052–3060. Available at: https://doi.org/10.1080/09638288.2018.1490461.

3. J. S. Johansen & C. Løvschall (2023). Region Midtjylland. Personal communication (November 12th, 2023).

4. Larsen, R.S. (2018) 'Rehabilitation and Climate Adaptation in One Place', Global Opportunity Explorer, 24 October. Available at: https://goexplorer.org/rehabilitation-and-climate-adaptation-in-one-place/ (Accessed: 18 December 2024).

Photos by Alfred Figueras Anton.

EU_03

Connecta Congost Natura 2025: Rehabilitating the Congost River

Context Overview

NBS TYPOLOGY	ECOSYSTEM TYPE
Restoration	Green spaces (gen.); Stormwater infrastructure; Rivers & waterways; Forests & trees
IMPACT SCALE	**URBAN TYPOLOGY**
City	Urban core
SOCIAL CHALLENGES ADDRESSED	**CO-BENEFITS**
Flooding	Increased biodiversity and their functions; Place regeneration & urban revitalization

With over 63,000 residents, Granollers near Barcelona, has historically developed along the Congost, which once sustained livelihoods through its water, building materials, and firewood. Rapid urban expansion before the 1950s led to the river's canalization, enabling residential and industrial development that turned away from the river (X. Romero, personal communication, 2023). Like many Mediterranean rivers, the Congost has highly variable flows—nearly dry in summer except for nutrient-rich wastewater, but prone to surging during spring and especially autumn downpours. These sudden increases, combined with altered river morphology, have often caused flooding, damaging infrastructure and endangering residents.

Introduction to the case study

CO-CO-NAT is a revitalization program for the Congost River and its surroundings in Granollers, one of the most populous cities in the industrial belt around Barcelona, Spain. With climate change significantly increasing the frequency and severity of flooding events, cities like Granollers need to adapt their urban river system. The severe deterioration of the ecological values of the river bed and adjacent habitats and the artificially-looking morphology caused by the concrete ribs containing the river bed has caused an overall public discontentment. Vandalism and the disposal of trash in the area compounded the discontentment. In sum, the river that was once the identity of the city and its residents has now deteriorated so much that it has become an unwanted element that is rejected by most residents (X. Romero, personal communication, 2023). The aim of the CO-CO-NAT 2025 project is to address the aforementioned social and ecological challenges and to transform the river into a fully functioning ecological and social asset for the city.

Nature-based Solutions characteristics

To achieve CO-CO-NAT, procedural aspects in the governance of this nature-based solutions project are as crucial as the final biophysical, landscape transformations. Concerning the biophysical transformations, the project can be zoned into three main intervention areas: the ecological rehabilitation of 300 m long and 21,000 m² in surface area section of the Congost River, the revitalization and re-naturalization of an adjacent traffic road and vehicle parking lot, turning it into a permeable promenade of 21,500 m², and the greening of two residential superblocks connected to the site (X. Romero, personal communication, 2023).

With these interventions, the municipality aims not only to enhance the river's ecological integrity and functions but also to update and regenerate the adjacent urban matrix, making it more accessible and attractive to the public. This urban renewal and ecological rehabilitation project is rooted in and partially implements the Natural Heritage Plan, the Strategic Plan 2030, and the Urban Greenery Plan (City of Granollers, 2022).

Governance

Participatory governance is key for the delivery of social co-benefits derived from the nature-based solutions decision-making process. CO-CO-NAT 2025 is tightly connected to, and makes use of the outcomes of the HORIZON 2020-funded INTERLACE project. More specifically, the INTERLACE Multi-Criteria Decision Analysis (MCDA) framework was used in co -creation workshops with local residents, practitioners, and researchers to collectively identify and prioritize criteria for each of the three nature-based solutions intervention areas.

The CO-CO–NAT project, launched in 2022, will transform nearly seven hectares of the city by 2025. This large-scale urban renewal and ecological restoration initiative, with a total budget of 4.2 million EUR (USD 4.4 million), has secured 3.2 million EUR (3.36 million USD) in funding from the Fundación Biodiversidad of the Ministry for the Ecological Transition and Demographic Challenge (MITECO), through the NextGeneration EU programme (City of Granollers, 2022). The programme constitutes the European Union's largest economic recovery plan, established in response to the COVID-19 pandemic (Gov. España, 2024). Important partners of the project include the Fundación Ecología Urbana y Territorial (FEUT), an independent research center that supported the City of Granollers in elaborating the project proposal. In addition, The ICTA-UAB is the research institution facilitating some of the participatory governance workshops.

Outcome

The project is currently in the implementation phase, with the first streets de-paved in early July 2024. Nevertheless, some insights can already be gleaned from this stage. Despite the utility of the INTERLACE MCDA approach in gathering residents' perspectives and expectations on the project, the results from these participatory workshops must be interpreted with caution. Notably, these workshops involved a relatively limited group of stakeholders, potentially skewing the representation of the city's socio-demographic reality. Some user groups may lack the interest or resources to participate fully, raising concerns about the fairness of outcomes, particularly in terms of justice and recognition.

To address this issue, future workshops related to the project will be held in alternative venues such as schools and public spaces, aiming to engage with harder-to-reach and minority groups such as youth, elderly people, and ethnic minorities. This approach seeks to mitigate power imbalances and ensure a more accurate representation of diverse needs and opinions. Additionally, a challenge lies in the subcontracted environmental engineering and landscape architecture firms' ability to incorporate insights from participatory workshops, which could impact overall transparency and trust in local government decisions.

Previous state (left) and rendered, future state of the river and adjacent park after the project is implemented (right)

Previous state (left) and rendered, future state of the re-naturalized streets connected to the river after the project is implemented (right)

Removal of paved surfaces and parking lots by the Congost river.

Writing: Oriol Garcia-Antúnez
Review: Veronica Olivotto
Copy Editing: Veronica Olivotto
Design: Phuong Anh Bui

Acknowledgements:

I would like to thank Sara Maestre Andrés and Johannes Langemeyer to involve me in one of the participatory workshops of the CO-CO-NAT 2025 project. I would also like to thank Xavier Romero from the City of Granollers for sharing his local knowledge and expertise on the project with me.

References:

1. City of Granollers (2022) Granollers obre el riu a la ciutat i crea una gran anella verda, Ajuntament de Granollers. Available at: https://www.granollers.cat/noticies/alcaldessa/congost (Accessed: 17 December 2024).

2. Govierno de España (2024) Plan de recuperación para Europa: NextGenerationEU, Plan de Recuperación, Transformación y Resiliencia. Available at: https://planderecuperacion.gob.es/plan-de-recuperacion-para-europa (Accessed: 17 December 2024).

3. X. Romero (2023). Environment and Nature technician, City of Granollers. Personal communication (March 30th, 2023).

Photos by The City of Granollers.

EU_04 Grønne Veje: Stormwater Management in Copenhagen

Context Overview

LOCATION

NBS TYPOLOGY Constructed & novel ecosystems	ECOSYSTEM TYPE Stormwater infrastructure; Green spaces (gen.)
IMPACT SCALE Greater urban region	URBAN TYPOLOGY Peri-urban
SOCIAL CHALLENGES ADDRESSED Flooding	CO-BENEFITS Recreational benefits

Copenhagen is the largest city and the capital of Denmark. With an extension of 90,10 square kilometers it is home to more than 650,000 inhabitants (Statistics Denmark 2023). Copenhagen is located in the eastern part of Denmark and it belongs geographically to the Øresund region, one of the most densely populated areas in Scandinavia. Copenhagen has an oceanic climate with frequent low-pressure systems from the Atlantic which bring abundant rainfall, especially from July to September. As in other European cities, future weather projections point to an increase in precipitation during winters and more frequent extreme rainfall events in summer. Simultaneously, the city is experiencing urban growth and densification, with some neighborhoods having experienced more than a 85 percent growth in the last 20 years (City of Copenhagen 2023). As a result, the paving of urban surfaces has increased at the expense of urban green spaces.

Introduction to the case study

Grønne veje (green street in English) is a street-based series of interventions to increase stormwater absorption in Copenhagen. With its flat topography and a combined sewage system, Copenhagen has taken steps to adapt to increasingly frequent extreme rainfall events like the ones in summers 2010 and 2011, which caused severe flooding and sewage overflow. Already in 2012, Copenhagen's cloudburst management plan recognized the need of implementing "green solutions" to mitigate flood risk and reduce the pressure in the combined sewage systems. Based on the plan the city designed and implemented "green streets" for on-site treatment of stormwater management over a 20 years period.

Nature-based Solutions characteristics

A *Grønne veje* consists of a traditional street where vegetated curve extensions with infiltration trenches are implemented to delay, retain and infiltrate stormwater on site. A curve extension consists of a soil mixture with high drainage capacity that purifies and infiltrates rainwater (HOFOR 2023). The vegetation chosen for the curb extensions consisted of herbaceous perennials and bulbs specifically adapted to extreme conditions (e.g., *Geranium, Echinacea, Hylotelephium, Tulipa, Narcissus* species). These plants were selected for their ability to withstand both periods of flooding and prolonged drought favoured by the soil's excellent drainage properties. The project intends to ultimately cover flooding-vulnerable areas, where green streets are technically feasible and socially accepted, focusing on streets that are privately owned by landowners association (*grundejerforeninger* in Danish), which have legal ownership of these spaces.

Governance

Through the Cloudburst Management Plan, the city of Copenhagen reserved 900 million DKK (147.5 million USD) towards green streets. This budget was assigned to HOFOR, the public water utility company in Copenhagen's metropolitan area, to develop and integrate these green solutions as water utility infrastructure. A significant portion of the budget allocated for the "green streets" initiative has been directed toward privately owned streets, highlighting an intriguing funding mechanism. This approach helps alleviate the financial burden on private owners, who would otherwise have to cover the costs of adapting their streets for stormwater management on their own. The "utility infrastructure" status under which these rain beds are implemented, however, severely restricts the budget for non-technical aspects, limiting their potential to enhance well-being, livability, and biodiversity. For instance, participatory processes and stronger citizen engagement in these biophysical transformations could have fostered a greater sense of ownership, strengthened local community ties, and improved perceptions of procedural and recognitional justice. However, HOFOR is legally prohibited from allocating funds for purposes beyond ensuring the technical functioning of this infrastructure, significantly constraining the delivery of broader social and environmental co-benefits. To break with the siloed administrative and funding structures, future projects like this should be more closely aligned with urban development and urban renewal plans to maximize its co-benefits and integration into other policy agendas such as biodiversity conservation and participatory governance (Lotte Kau Andersen, personal information 2023).

Outcome

The pilot phase of the project, which started in 2018, ended in 2022. As of 2024, "green streets"— featuring diverse designs and sizes—have been implemented in 23 locations, primarily in low-density suburban neighborhoods in and around Copenhagen. These streets have been re-greened with rain beds, which add aesthetic value as well as a traffic pacifying element very much appreciated by local residents. Given the lack of previous experience in developing nature-based solutions for stormwater management, this piloting phase has been the testing ground for HOFOR to gain experience and develop a more cost-effective prototype of rain beds that will be upscaled to meet the goals of the Cloudburst management plan. An important innovation of this project is the integration of these rain beds as technical infrastructure under the domain and responsibility of HOFOR. This transfer of responsibility from the landowner association to HOFOR requires the latter to ensure the well-functioning and correct maintenance of the rain beds as an integral part of the water utility infrastructure, which secures their long term sustainability and success free of charge for local residents. Yet, this has come with some challenges. First, the implementation of these rain beds in privately owned streets requires the final consent from the board landowners' association, whose attitudes might change over the period of negotiations. Due to the space these rain beds take, both below and above ground, their implementation might not be technically possible (e.g. due to competing underground infrastructure) or socially unacceptable (e.g., due to loss of parking space and bike infrastructure). Therefore, HOFOR has focused on low-density areas, acknowledging the difficulties of implementing them in the city center.

Finalized piloting site in Valby nye Villakvarter (Copenhagen)

Finalized piloting site in Valby nye Villakvarter (Copenhagen)

Writing: Oriol Garcia-Antúnez
Review: Veronica Olivotto
Copy Editing: Veronica Olivotto
Design: Phuong Anh Bui

References:

1. City of Copenhagen (2023) Nøgletalsanalyser, Københavns Kommune. Available at: https://www.kk.dk/om-kommunen/fakta-og-statistik/noegletalsanalyser (Accessed: 1 October 2023).

2. HOFOR (2023) Grønne veje—veje med regnbede, HOFOR. Available at: https://www.hofor.dk/baeredygtige-byer/udviklingsprojekter/skybrudssikring/groenne-veje/ (Accessed: 1 October 2023).

3. Lotte Kau Andersen (2023), HOFOR, personal communication

4. Oriol Garcia Antunez (2023) Demonstration video: https://youtube.com/shorts/a3-k17V1cQI?feature=share

5. Statistics Denmark (2003) 'Befolkningen 1. januar efter byområder, landdistrikter, alder og køn'. Available at: https://statistikbanken.dk/statbank5a/SelectVarVal/Define.asp?MainTable=BY1 (Accessed: 1 October 2023).

Photos by Oriol Gracia-Antúnez

EU_05 UNaLab Tampere: Living Labs for Nature-Based Solutions

Context Overview

LOCATION

NBS TYPOLOGY Restoration; Constructed & novel ecosystems	ECOSYSTEM TYPE Green spaces (gen.); Roofs & walls; Stormwater infrastructure; Forests & trees
IMPACT SCALE Neighbourhood/Community	URBAN TYPOLOGY Urban core
SOCIAL CHALLENGES ADDRESSED Flooding; Biodiversity loss; Water quality	CO-BENEFITS Recreational benefits; Stronger human-nature connections

Tampere is the third largest city in Finland and the largest inland center in the Nordic countries with 260,319 inhabitants. The city is continuously expanding, attracting more people from the metropolitan area as a hub for technology, research, education, culture, sports, and business. Due to the rapid population growth, Tampere confronts significant challenges, such as water pollution and reduced biodiversity. Moreover, heavy rainfall, flooding and rising volumes of water are expected to pour over the Tampere region, with an increase between five percent (RCP2.6) and 15 percent (RCP8.5) by 2080 (Gregow et al, 2021).

Introduction to the case study

The UNaLab (Urban Nature Labs) project aims to create more sustainable, resilient, and inclusive cities by co-creating and testing nature-based solutions in real-life settings through living labs—user-driven innovation spaces where researchers, citizens, and other stakeholders collaborate in everyday urban environments. Vuores and Hiedanranta have been UNaLab's primary living lab sites in Tampere for nature-based solutions implementation. Nature-based solutions in this case are intended as knowledge and practices to deliver multiple economic, social and environmental benefits and co-benefits. Vuores, a former greenfield, is a residential area established in 2007, while Hiedanranta is a brownfield (former pulp mill area) currently unpopulated, with municipal plans to include housing for 25,000 residents over the next 30 years. In the Vuores district the main challenges addressed have been flood prevention and stormwater management. Another challenge taken up by both sites has been increasing the water quality of stormwater. Both sites followed co-creation processes to enhance awareness and mutual learning between different stakeholders. Co-creation consisted in a series of workshops for experts, residents and urban planners, alongside school children to build knowledge and consensus around nature-based solutions implementation. Moreover, some workshops aimed at envisioning, ideating and implementing/testing nature-based solutions in the study areas.

Nature-based Solutions characteristics

The following nature-based solutions were implemented in Vuores: retention basins, alluvial meadows, urban gardens with small-scale nature-based solutions and wooden boardwalks (duckboards) to improve the citizens' access to natural areas (Koukkujärvi nature trail). Through these interventions, the water running through the system is purified before it ends up in the lakes surrounding Vuores. Hiedanranta had smaller interventions focused on wastewater treatment, such as biofilters and microalgae systems, where algae grow using nutrients found in human urine and can later be used as biomass for bioethanol production. Moreover, a green roof and wall have been installed in the city center area. The City of Tampere, partner of the project, launched in 2019 a call to housing cooperatives and associations in Vuores for small-scale nature-based solutions, offering three innovation vouchers of up to 10,000 EUR (11,300 USD) (Särkilahti et al., 2019). Two of them were used to develop garden areas near residential housing, the third one made possible to create a community horse park in the Vuores Central Park.

Governance

The UNaLab project was financed by Horizon2020 and it lasted 5 years, from June 2017 to November 2022. The whole project cost 14.2 million EUR (16.8 million USD), receiving 12.7 million EUR (12.7 million USD) from the EU. VTT Technical Research Centre of Finland led UNaLab with an EU contribution of 1.1 million EUR (1.3 million USD). The EU contribution to the Tampere case consisted of 1.4 million EUR (1.7 million USD) to the City of Tampere and 186,375 EUR (219,922 USD) to the architecture, engineering, and consulting company Ramboll Finland. The Council of Tampere Region is a major actor in terms of climate adaptation planning: in 2018, Tampere city region approved a carbon neutrality by 2030 roadmap, with sustainable urban nature as one of the six themes. In the Tampere Region climate work is being carried out through the active cooperation between the regional government and the Tampere Region ELYkeskus (Centre for Economic Development, Transport and the Environment). Tampere region is part of HINKUmaakunnat, 'Low Carbon Regions', aiming to achieve an 80 percent reduction in regional greenhouse gas emissions by 2030, compared to the 2007 levels. The Regional Council has a representative in the National Adaptation Monitoring Group (Gregow et al., 2021).

Outcome

In May and June 2020, a survey was conducted among Vuores residents to assess the impact of local nature-based solutions on their quality of life (UNaLab, 2020). The findings revealed that the stormwater parks, Koukkujärvi nature trail, and community horse park were widely regarded as safe and recreationally valuable, offering activities such as birdwatching and enjoying scenic views. Notably, the nature trail and stormwater parks provided residents with relaxation and a sense of closeness to nature. Additionally, 43 percent of respondents reported gaining new knowledge about nature and ecosystems through informational signs installed in the lake Koukkujärvi in spring 2019. At the start, there were some challenges. For example, it was difficult to find enough space to build the biofilters in Virolainen Park. Some residents also thought the wetland areas and meadows didn't look very appealing. To address these concerns, the city organized workshops and put up signs to explain the purpose and value of nature-based solutions. These efforts helped people better understand the ecological benefits. In fact, the feedback from these workshops led the city to make changes to its plans for Vuores, giving more attention to biodiversity, water management, and public well-being (Särkilahti et al., 2019).

Enrichment meadows implemented in Tampere

Biofilters implemented in Tampere

Graffiti close to the implementation site in Tampere

Writing: Eugenia Castellazzi
Review: Veronica Olivotto
Copy Editing: Veronica Olivotto
Design: Phuong Anh Bui

References:

1. Gregow, H. et al. (2021) 'Ilmastonmuutokseen sopeutumisen ohjauskeinot, kustannukset ja alueelliset ulottuvuudet', Suomen ilmastopaneeli, p. 166. Available at: https://helda.helsinki.fi/items/d788de23-a628-4631-a66e-de63e08893f2 (Accessed: 13 December 2024).

2. Sarabi, S. et al. (2021) 'Barriers to the Adoption of Urban Living Labs for NbS Implementation: A Systemic Perspective', Sustainability, 13(23), p. 13276. Available at: https://doi.org/10.3390/su132313276.

3. Särkilahti, M., Mustajärvi, K. and Leppänen, S. (2019) 'Biosuodattimia, hevoshakoja ja viherkattoja – yhdessä oppiminen luontopohjaisten ratkaisujen luomisessa', Alue ja Ympäristö, 48(2), pp. 20–37. Available at: https://doi.org/10.30663/ay.85117.

4. Särkilahti, Maarit (2019) 'Co-creating nature-based solutions in EU project demonstration city Tampere', Rakennustekniikka.

Photos by Eugenia Castellazzi

EU_06 UPSURGE: Soil Remediation and Urban Agriculture in Belfast

Context Overview

NBS TYPOLOGY	ECOSYSTEM TYPE
Restoration	Agriculture & food spaces
IMPACT SCALE	**URBAN TYPOLOGY**
Site	Urban core
SOCIAL CHALLENGES ADDRESSED	**CO-BENEFITS**
Social inclusion; Land degradation	Soil stability, fertility & quality; Recreational benefits; Place regeneration & urban revitalization

Belfast, the capital and largest city of Northern Ireland, has a growing population of 345,000 and is among the youngest urban centres in Europe (Northern Ireland Statistics and Research Agency, 2021). Located on the eastern coast at the mouth of Belfast Lough, the city is known for its industrial heritage, including linen production and the Titanic. Decades of conflict reinforced social and physical divisions, and while peace-building efforts continue, challenges such as segregation, poverty, poor mental health, and inequality persist. Environmental issues include flooding, extreme weather, and poor air quality (Belfast City Council, n.d.).

Introduction to the case study

UPSURGE is a nature-based solutions demonstration site in Belfast that combines soil remediation with urban agroecology through co-production. A large, underused green space owned by Belfast City Council (BCC) was chosen as the UPSURGE nature-based solution demonstration site. This site is a part of Belfast's Botanic Gardens and has been used as allotments and for sports and cultural events. Soil testing revealed widespread contamination linked to waste from previous industrial activities. Today, the site is popular with dog-walkers, and is well situated next to university facilities, local schools, a theatre, and several neighbourhoods. Before the project began, BCC carried out a consultation to determine the types of nature-based solutions to be implemented on the site. This informed the focus on sustainable urban agroecology and soil remediation to build social cohesion, foster nature education and environmental justice for a diverse mix of stakeholders across communities.

Nature-based Solutions characteristics

UPSURGE requires project partners to use co-creation processes to involve citizens, academia, local governments, environment groups and businesses throughout the project. Once the project started in Belfast, BCC and Queen's University Belfast (QUB) teamed up to plan co-design activities to determine the exact brief and form of nature-based solutions on the site. From April to September 2022, this team hosted three half-day co-design workshops. The first workshop explored the scales of city, neighbourhood, and demonstration sites to consider current perceptions of nature, existing routes, important stakeholders, and potential nature-based solutions. The second event looked at themes of green, water and mobility to explore possible interventions at different scales, from new green corridors to community gardens and nature engagement events. The last workshop presented a concept for a range of nature-based solutions on the site. Stakeholders with similar aims were grouped to critique the concept and collectively decide priorities for the project. After the co-design workshops, QUB created a nature-based solutions design guide for the landscape architect nominated to carry out the detailed design. At the same time, community group, Friends of the Field, hosted events on the site including a BioBlitz to log existing plant and insect species. The co-design process led to the development of a range of nature-based solutions interventions on the site. QUB firmed up plans for a research garden for soil remediation and carbon testing. Friends of the Field set out an intention to set up and run a community garden with education activities and food production. Another group, Friend of Botanic Gardens, proposed a horticultural garden to test climate-adaptive plant species. A new site entrance and pathways were also proposed to support inclusive and sustainable travel to and through the site.

Governance

The UPSURGE project received funding from the Europe Union's Horizon 2020 research and innovation program under grant agreement No. 101003818. The project has 23 partners and runs for four years, from September 2021 to August 2025. The UPSURGE project supports five European city partners (Belfast, Maribor, Breda, Budapest and Katowice) to test nature-based solutions on urban sites to address urban regeneration, improve air quality and verify outcomes through a localised sensor network. Each city chose different kinds of nature-based solutions, reflecting specific climate and social challenges. BCC received 660,375 EUR (693,393 USD) to plan, manage, and fund the implementation of nature-based solutions in Belfast. QUB received 548,550 EUR (575,977 USD), with part of this funding used to monitor and support all five cities to co-design, construct and deploy nature-based solutions on urban demonstration sites.

Outcome

The co-design process enabled a wide range of people to meet and contribute to planning the future of Lower Botanic Gardens and other satellite sites in the city (Campbell et al., 2023). Workshops created ideal spaces to test the feasibility of different nature-based solutions. They also highlighted the need for ongoing communication and engagement across and beyond the project (McKinley et al., 2023). Negotiations on the governance of the community garden, coupled with extreme weather events have caused delays in implementing nature-based solutions in Belfast. Despite this, construction has begun. The research gardens are operational, and new pathways, fencing and hedging have been implemented. With support from QUB and BCC, Friends of the Field and The Conservation Volunteers have taken the lead to agree on the layout of the community garden which should be constructed in Spring 2024, partly by landscape contractors and partly by community volunteers.

Implementing QUB's research gardens

Capturing a cross-pollination of design ideas at the site scale in the second co-design workshop

Writing: Emma Campbell
Review: Veronica Olivotto
Copy Editing: Veronica Olivotto
Design: Phuong Anh Bui

Acknowledgements:

The co-design process was developed and led collaboratively, with great effort from others including Professor Jennifer McKinley, Professor Greg Keeffe, Dr Bakul Budhiraja, Dr Siobhan Cox, Dr Sean Cullen and Jennifer Newell at QUB, and Richard McLernon, Mura Quigley, and Alan Wardle at BCC. Further recognition is owed to key stakeholder groups as well as UPSURGE co-creation project partners, Dr Janine Maier, and Christina Haderer.

The content of this case study reflects the views of the author only. Neither the Research Executive Agency nor the European Commission are responsible for any use that may be made of the information contained herein. This document contains information, which is proprietary to the UPSURGE consortium. Neither this document nor the information contained herein shall be used, duplicated or communicated by any means to any third party, in whole or in parts, except with the prior written consent of the UPSURGE consortium. This restriction legend shall not be altered or obliterated on or from this document. Neither the Research Executive Agency, the European Commission nor the UPSURGE project consortium are liable for any use that may be made of the information that it contains.

References:

1. Belfast City Council (no date) Resilient Belfast, Belfast City Council. Available at: https://www.belfastcity.gov.uk/Business-and-investment/Resilient-Belfast (Accessed: 16 July 2025).

2. Campbell, E. et al. (2023) 'Lessons on co-designing nature-based solutions for urban regeneration and pollution alleviation in the EU Horizon "UPSURGE" project: QUB Post-Doctoral Centre Post-doc Showcase 2023', in.

3. McKinley, J. et al. (2023) Co-designed infrastructural demonstration plans. UPSURGE Project.

4. Northern Ireland Statistics and Research Agency (2021) Census. Available at: https://www.nisra.gov.uk/statistics/census/census-2021 (Accessed: 16 July 2025).

5. UPSURGE (no date) 'Nature Based Solutions & Cities', UPSURGE Project. Available at: https://www.upsurge-project.eu/nature-based-solutions-cities/ (Accessed: 17 December 2024).

Photos by authors

EU_07 UNaLab Eindhoven: Living Labs for Nature-Based Solutions

Context Overview

NBS TYPOLOGY Restoration; Constructed & novel ecosystems	**ECOSYSTEM TYPE** Green spaces (gen.); Roofs & walls; Corridors; Rivers & waterways
IMPACT SCALE City	**URBAN TYPOLOGY** Urban core
SOCIAL CHALLENGES ADDRESSED Heat; Flooding	**CO-BENEFITS** Urban cooling; Increased biodiversity and their functions

Eindhoven is one of the five biggest cities in the Netherlands, with around 230,000 people living there. The wider region has more than 750,000 residents. It plays an important role in the country's economy, making up about 14 percent of the national income. Located in the Brainport region, a hub for high-tech industries, innovation, and design, Eindhoven is known for its creativity and knowledge-based economy. But the city is facing growing challenges. Its population is expected to rise from 220,000 in 2014 to around 300,000 by 2030. Climate change is making some problems worse, such as flooding, extreme heat in the summer, air pollution, and the loss of natural areas like rivers, ponds, and green spaces that help keep the city cool and healthy. These changes are putting pressure on the urban environment and affecting people's well-being.

Introduction to the case study

UNaLab utilizes a living lab methodology to bring nature-based solutions to the central streets of Eindhoven. The City of Eindhoven decided to use nature-based solutions in several places, mainly in the city centre. The goal of this project is to better connect water features (like canals and ponds), green spaces (such as trees and parks), and buildings, so that the city becomes a better place for people to live. In the UNaLab project, nature-based solutions means using natural methods and knowledge to bring economic, social, and environmental benefits at the same time. Many of these nature-based improvements were added to projects that were already being planned or built. By using a Living Lab method, UNaLab activities gave the city a chance to include more nature into those developments.

Nature-based Solutions characteristics

Eindhoven has introduced several nature-based solutions in its city centre. In many streets that used to be fully paved, trees and greenery have been added. These changes help manage rainwater, reduce heat, and create more pleasant spaces for people. On Vestdijk, a major road in the centre, car traffic has been reduced, and existing green areas have been improved—some now include spaces designed to hold excess rainwater. In Clausplein square, trees, shrubs, and grasses have been planted with the goal of soaking up and cleaning rainwater. In the old industrial area known as NRE, where the historic buildings are now protected monuments, new green walls and plantings—including trees, flowering plants, and grasses—have helped bring life back to the area. Other examples of greening include, a green wall on a building along Oude Stadsgracht, and a green roof on City Hall. The entrance to City Hall has also been updated with green spaces planted with native species. Outside the city centre, a section of the Grote Beek stream—once partly removed—has been restored. This has reconnected parts of the water system, improved water quality, and helped wildlife move more freely, supporting local biodiversity.

Governance

The UNaLab project was financed by Horizon2020 and it lasted 5 years, from June 2017 to November 2022. The whole project cost 14.2 million EUR (16.1 million USD) and it received 12.7 million EUR (14.4 million USD) from the EU. The EU contribution to the Eindhoven lab consisted of 1.7 EUR (1.9 million USD) to the City of Eindhoven, 677,462 EUR (765,532 USD) to Eindhoven University of Technology, 232,120 EUR (262,295 USD) to the energy company Innohub Bv, 9,275.00 EUR (10,480 USD), to P.G. Kuijpers & Zonen Bv, an interim agency specialized in building and maintaining technical installations. The last company had several subcontractors for a total expense of 852,600 EUR (963,438 USD). A big part of the project is working closely with local people and organisations. This approach, called co-creation, is very important in Eindhoven—not just because Dutch law requires public participation, but also because many different groups are genuinely interested in being involved (Sarabi et al., 2021). Both the municipality and many residents were already experienced in co-creation processes. The co-creation meetings in spring 2018 led to the creation of a Community of Practice for nature-based solutions, involving many stakeholders, mainly relevant professionals from the municipality and industry, that participated in workshops and in the shaping of Living Labs (Laikari et al., 2021).

Outcome

Eindhoven has added greenery to many areas that were previously covered in concrete or pavement. These changes have brought nature closer to people's everyday lives. The nature-based solutions chosen by the city have brought several benefits: they help keep people healthy, provide shade and cooling during hot weather, reduce the risk of flooding, and support biodiversity—especially around the restored Grote Beek stream. Eindhoven University of Technology also held a workshop with the city's Community of Practice to explore challenges in using nature-based solutions in an urban living lab setting. The main barriers identified included: too many separate institutions involved (which made coordination difficult), not enough clear information for citizens about how nature-based solutions work or what they do, and a lack of effective ways to involve a broader and more diverse group of residents—considering differences in age, income, and gender. One of the most challenging projects was the attempt to "daylight" a section of a buried river—that is, to uncover and restore it. Because of disagreements during planning, the original location couldn't be used. This led to delays, and the project was eventually moved to another site that, while not a public space, remains open and accessible to people. Another difficulty came up in Clausplein square. The city had planned to install LiDAR cameras to study how the green spaces were benefiting people socially. However, this couldn't go ahead because of national privacy laws (Sarabi et al., 2021).

Writing: Eugenia Castellazzi
Review: Veronica Olivotto
Copy Editing: Veronica Olivotto
Design: Phuong Anh Bui

Vestdijk Boulevard

References:

1. Laikari, A. et al. (2021) NBS Demonstration Site Start-Up Report. Urban Nature Labs.

2. Sarabi, S. et al. (2021) 'Barriers to the Adoption of Urban Living Labs for NBS Implementation: A Systemic Perspective', Sustainability, 13(23), p. 13276. Available at: https://doi.org/10.3390/su132313276.

3. UNaLab (2022) NBS Eindhoven, ArcGIS StoryMaps. Available at: https://storymaps.arcgis.com/stories/dd013aeab97240398f0bb5917643c9e5 (Accessed: 13 December 2024).

Photos by Network Nature.
https://networknature.eu/casestudy/29189

EU_08 ProGIreg Living Lab for Green Infrastructure and Post-Industrial Regeneration

Context Overview

NBS TYPOLOGY Restoration; Constructed & novel ecosystems	ECOSYSTEM TYPE Agriculture & food spaces; Green spaces (gen.); Roofs & walls; Corridors; Forests & trees
IMPACT SCALE Site	URBAN TYPOLOGY Urban core
SOCIAL CHALLENGES ADDRESSED Nature connection; Physical and mental health; Heat	CO-BENEFITS Recreational benefits; Flood mitigation

Zagreb is the capital and largest city of Croatia. It is located in the north of the country, along the Sava River, at the southern slopes of the Medvednica mountain, near the border of Croatia and Slovenia. The Municipality of Zagreb holds 770,000 people, and its Metropolitan Area holds slightly above one million people. Zagreb has an oceanic climate with continental influences, with cold winters (average daily mean temperature around 1°C) and warm to hot and very wet summers (average temperature 22°C), with occasional short-lived heatwaves. The Living Lab developed by the ProGIreg project is located within the Sesvete district in the east of Zagreb. Sesvete has 70,000 inhabitants, and the youngest population in Croatia. The Living Lab is located at the site of a former meat-processing factory, Sljeme.

Introduction to the case study

ProGIreg is a Living Lab with the aim of implementing green infrastructure and better public spaces through residents participation. The EU funded the proGIreg project ('productive Green Infrastructure for post-industrial urban regeneration: nature for renewal) to improve the resilience of Sesvete to increasing climate-induced heat through urban regeneration with and for citizens, with an interest in fostering urban consciousness and identity. The main problems addressed included the lack of public green spaces, and the need for better public facilities, bike lanes, a hub with co-workspaces and a secure crossing of the railway.

Nature-based Solutions characteristics

The Living Lab implemented community-based urban gardens, aquaponics, green roofs and walls, and green corridors. The community-based urban farms and gardens were located in city-owned land upgraded with solar water distribution pumps and solar water purifying devices using renewable energy sources. Near the gardens, an info center with lectures, workshops, exhibitions, and round tables for the community was set up. The green walls, roofs and aquaponics were built by developing a modular urban farm that combined both nature-based solutions. The new accessible green corridors (cycling and pedestrian paths) were built on reused derelict land. In addition, ProGIreg implemented a therapeutic garden designed in collaboration with local associations working with people with disabilities, including spaces for activities and urban farming, and promoted tree planting activities, with 500 trees and shrubs planted by 250 citizens. Finally, it introduced low-carbon guidelines into new strategic documents.

Governance

The project proGIreg (2018-2023) was funded by the European Commission under the Horizon 2020 programme (1.1 million EUR; 1.3 million USD). In Zagreb, the Office for Economy, Environmental Sustainability and Strategic Planning was the coordinator of the Living Lab. Other local partners included a local NGO (Green and Blue Sesvete) and the City of Zagreb Bureau for Physical Planning and Zagreb Faculty of Architecture. Private companies created specific parts of the nature-based solutions. Local stakeholders were involved to ensure users and residents accepted and embraced the new nature-based solutions. These co-design processes (including workshops) were crucial to the successful implementation of the solutions, as well as, in the engagement of the local population in decision-making processes, allowing to detect underlying problems and risks. With this project, Zagreb established new protocols and changes to planning procedures and policy development processes for nature-based solutions.

Outcome

The project ProGIreg developed self-sustaining business models for nature-based solutions, based on scientific assessment of social, ecological, and economic benefits while changing locals' perception of the neglected South of the district. By creating productive green infrastructure, this project helped to improve living conditions and reduce vulnerability to climate change, while having measurable economic benefits to citizens and entrepreneurs.

The ecosystem services delivered include cultural services (recreation, social cohesion, cultural heritage, mental well-being, and health), and regulating services, namely the mitigation of the urban heat island effect. In specific, the co-benefits of the community urban gardens included the inclusion of vulnerable groups, namely people with disabilities who are now able to engage in gardening activities. The therapeutic garden improved the well-being of users by providing spaces for social interaction and contemplation. The green walls/roofs and aquaponics created economic value due to both the installation of an improved water distribution network and a solar purification device, as well as the creation of a business model related to the installation of aquaponic systems. Another outcome was the establishment of protocols and procedures for environmental compensation, namely the proposal of guidelines for low-carbon development planning, created following European green policies. Some challenges of this project included external factors (i.e. damages by an earthquake in 2020, land ownership issues) and procedural reasons which led to delays and cancellations of original plans for implementation.

Writing: Filipa Grilo
Review: Veronica Olivotto
Copy Editing: Veronica Olivotto
Design: Phuong Anh Bui

References:

1. proGIreg (2020) Deliverable 3.5 Implemented Living Lab in Zagreb, Croatia. GRAD ZAGREB.

2. proGIreg (no date) The Living Lab: Sesvete, proGIreg. Available at: https://progireg.eu/zagreb/ (Accessed: 26 January 2024).

Photo by City of Zagreb. https://progireg.eu/zagreb/

Green wall as part of the modular urban farm

EU_09 UrbanGreenUp: Green Infrastructure in Liverpool

Context Overview

NBS TYPOLOGY	ECOSYSTEM TYPE
Restoration; Constructed & novel ecosystems	Green spaces (gen.); Corridors; Forests & trees; Stormwater infrastructure; Roofs & walls; Productive blue spaces; Agriculture & food spaces
IMPACT SCALE	**URBAN TYPOLOGY**
City	Urban core
SOCIAL CHALLENGES ADDRESSED	**CO-BENEFITS**
Heat; Flooding; Water quality; Nature connection;	Increased biodiversity and their functions; Carbon sequestration; Recreational benefits; Green jobs and economic benefits

Liverpool is a city and metropolitan borough in Merseyside, North West England, located on the eastern side of the Mersey Estuary, adjacent to the Irish Sea. The city holds around 485,000 citizens and its City Region has around 1,550,000 people. The wider Liverpool area is the third largest in England and Wales outside London. Liverpool is characterized by a temperate maritime climate, with relatively mild summers, cool winters and rainfall spread evenly throughout the year. In 2008, Liverpool was nominated the annual European Capital of Culture. Its local economy is based on the service sector, being one of the largest in the United Kingdom.

Introduction to the case study

URBAN GreenUP introduces a variety of green infrastructure projects to improve soil conditions, biodiversity, cooling and rainwater storage in Liverpool. Liverpool has faced challenges in managing its urban form to pursue economic growth, leading to the conversion of green space into urban land, affecting connectivity, ecosystem services delivery, and climatic resilience. The city nowadays supports various initiatives aiming to develop and implement green infrastructure and renature urban plans, having joined the "Liverpool City Green Infrastructure Strategy". Specifically, there has been a need to tackle urban challenges related to poor environmental quality, lack of green spaces, and lack of connectivity, which threaten human health, quality of life, and urban biodiversity. In addition, there has been a need for strategies that promote nature connection and interactivity that also encourage investments to build a city attractive to live and work, increase social justice and social cohesion.

Nature-based Solutions characteristics

During the URBANGreenUp project, 40 types of nature-based solutions were implemented to re-naturalize urban areas. These solutions included the construction of green corridors with pedestrian and cycle lanes, planting of trees and shrubs for climate heat mitigation, carbon capture, and air quality improvement, sustainable drainage systems and rain gardens. Green walls and roofs, and mobile and floating gardens were installed in areas lacking spaces for large green spaces. For the improvement of soil functionality and biodiversity, insect hotels and bird and bat boxes were built.

Governance

The URBAN GreenUP (2017-2022) was a project funded under the European Union's Horizon 2020 programme (14 million EUR in total; 16.5 million USD). Its objective was the development, application and replication of Renaturing Urban Plans in European and non-European partner cities to mitigate the effects of climate change, improve air quality and water management, and increase urban sustainability through innovative nature-based solutions. The URBAN GreenUP consortium comprised eight partner cities, divided into front-runners (three European cities, including Liverpool), the followers, and the Network of Cities. The project coordinator was the CARTIF Technology Center. The project aimed to help make a step change in the quantity and quality of green infrastructure across the city, the wider city region and Mersey Forest. The planting of trees across multiple city sites included the work of a wide range of stakeholders and landowners.

Outcome

With the introduction of green infrastructure (particularly tree planting), GreenUp was able to reduce carbon emissions, affecting 160,000 citizens per year, and reduce temperature (2-4°C during summer), organic gas evaporative emissions, CO_2, and NO_x emissions, affecting 30,000 citizens. In addition, the new nature-based solutions improved accessibility, connectivity to existing green infrastructure, and air quality, serving as pollutant filters. The water interventions reduced flood risk and contributed to rainwater storage, improved water quality, and provided ecological benefits concerning the surface water drainage network. There was an overall increase in urban biodiversity levels, with a particular increase in plant species richness and functional diversity of plants and insects, and insectivore abundance, leading to higher use of corridors for movement, which has been attracting many visitors.

The new nature-based solutions provide co-benefits related to social justice and social cohesion, namely crime reduction and improved safety on green routes, accessible green spaces, job creation, and opportunities for social interactions. In addition, these solutions have contributed to changes in house prices, savings in energy use, improvements in citizens' health and well-being, and new options for educational activities. Finally, this project has improved the perceptions of citizens on urban nature and green spaces quality, increased social learning concerning nature-based solutions and improved citizens' engagement with nature.

Writing: Filipa Grilo
Review: Veronica Olivotto
Copy Editing: Veronica Olivotto
Design: Phuong Anh Bui

References:

1. Urban GreenUp (no date a) Deliverables, Urban GreenUp. Available at: https://www.urbangreenup.eu/resources/deliverables/ (Accessed: 17 December 2024).

2. Urban GreenUp (no date b) Liverpool, Urban GreenUp. Available at: https://www.urbangreenup.eu/cities/front-runners/liverpool.kl (Accessed: 28 January 2024).

Photo by Dr. Juliet Staples, Liverpool City Council.

Wapping dock-floating ecosystem island installed during the URBAN GreenUP project

EU_10

CLEVER Cities: Urban Greening and Co-design in Milan

Context Overview

NBS TYPOLOGY Constructed & Novel Ecosystems	ECOSYSTEM TYPE Roofs & walls; Green spaces (gen.)
IMPACT SCALE City	URBAN TYPOLOGY Urban core
SOCIAL CHALLENGES ADDRESSED Heat; Biodiversity loss; Flooding	CO-BENEFITS Recreational benefits

The CLEVER Cities project focused on implementing nature-based solutions within Milan's urban renewal strategies, while also promoting biodiversity and sustainability. Milan is the main city of the Lombardy region and serves as a major administrative, economic, and cultural center in Italy, with a population of approximately five million.

Milan faces ongoing urban challenges. High levels of air pollution, aggravated by increasing urbanization, pose significant risks to public health. Moreover, over 60 percent of Milan's land is covered in impermeable surfaces, making it one of the most paved cities in Italy. This limits space for greenery and heightens vulnerability to climate impacts.

Introduction to the case study

The city is an active member of several international networks, such as Eurocities, C40 Cities Climate Leadership Group, the Urban Development Network, and 100 Resilient Cities, reflecting its strong commitment to sustainable urban development and climate resilience.

In recent years, Milan has transformed several abandoned areas within its metropolitan region. These projects have provided the city with valuable opportunities to create new green spaces while revitalizing entire neighborhoods. They have also showcased the potential for integrating nature-based solutions into broader urban regeneration efforts. CLEVER Cities was a collaborative initiative involving three European cities—Hamburg, London, and Milan—aimed at positioning the EU as a global leader in nature-based innovation. The project employed a city-focused model, addressing urban regeneration challenges and collaborating with local partners to foster inclusive, sustainable transformation. In Milan, the project focused on the city's southern districts and was implemented through three "Clever Action Labs" (CALs). Each CAL had a specific focus:

CAL 1 – Green Roofs and Facades: Aimed at raising awareness of the benefits of green roofs and facades while supporting the installation of pilot projects. This lab promoted innovation in retrofitting buildings with green features, combining technical expertise with community input to ensure social acceptance and aesthetic integration.

CAL 2 – Co-design of Public Green Spaces (Giambellino 129): Focused on transforming various public areas—such as urban gardens, bird habitats, and vertical gardens—through participatory planning. Residents and community organizations were directly involved in designing and managing these spaces, fostering a sense of ownership and ensuring long-term care.

CAL 3 – Greening of Railway Infrastructure: Explored ways to integrate green elements into Milan's southern railway corridors. The lab showcased how nature-based solutions can be applied even in areas dominated by industrial and transport activities, enhancing their ecological value and visual appeal.

Each lab served as a platform for testing nature-based interventions and engaging communities in the regeneration process, aiming to embed sustainability into daily urban life.

Governance

The CLEVER Cities project in Milan had a total budget of 14.8 million EUR (17.5 million USD), with 14.2 million EUR (16.7 million USD) contributed by the European Union. The project ran for six years, from June 1, 2018, to November 30, 2023. The governance model combined institutional leadership with inclusive participation. A wide range of local organizations participated, including RFI (Italy's rail infrastructure company), Politecnico di Milano, Ambiente Italia (an environmental consultancy), Eliante (a social cooperative), and WWF. The Milan case study focused on co-creating locally tailored nature-based solutions to achieve tangible environmental, social, and economic benefits. One of the project's primary objectives was to embed co-creation into mainstream urban planning practices. For instance, the newly approved Piano Territoriale Metropolitano (PTM) acknowledges nature-based solutions as a strategic tool for climate adaptation and highlights collaboration with local stakeholders, especially through Public-Private Partnerships (PPPs).

Outcome

The primary outcomes of the CLEVER Cities project in Milan are closely linked to the co-creation process at the initiative's core. A key insight from the project was that aligning co-creation with financial incentives can greatly enhance public engagement. Thus, developing tools and procedures that integrate public participation and co-design into the criteria for financial support proved to be a vital step toward mainstreaming these practices.

In the case of CAL 1, which focused on green roofs and facades, the project emphasized the importance of combining local residents' insights with expert input during the co-design process. This collaborative approach resulted in solutions that were both socially accepted and technically sound for retrofitting private buildings with green infrastructure.

Another important lesson emerged regarding governance in multi-property buildings. Shared governance mechanisms were crucial for accelerating environmental upgrades in alignment with climate neutrality goals for 2030 and 2050. These mechanisms facilitated cooperation among residents, ensuring that decisions about building improvements, such as installing green roofs, were collectively supported and maintained over time.

The project also highlighted the need for more complex financial models in working with citizens. Rather than relying exclusively on profit-driven frameworks, the economic feasibility of many nature-based solutions was evaluated based on avoided costs—such as reduced flooding risks and lower energy consumption. Communicating these less-visible financial benefits was a crucial strategy, as it helped expand understanding of the various advantages nature-based solutions can provide, including their social and environmental co-benefits. This, in turn, emphasized the importance of assessing nature-based solutions through a framework of long-term social return on investment, rather than merely immediate monetary gain.

Despite these successes, the project also identified key challenges. In Milan, the lack of a deeply rooted culture of collaboration in decision-making, along with the absence of well-established policies for cooperative planning, hindered more fluid engagement processes. Addressing these barriers required building trust between institutions and the public, providing accessible entry points for participation, and ensuring that co-creation processes remained flexible and adaptable. The project demonstrated that for nature-based solutions to be fully effective, the public must feel genuinely involved and empowered in shaping the urban spaces they inhabit.

Writing: Jarumi Kato Huerta
Review: Veronica Olivotto
Copy Editing: Veronica Olivotto
Design: Phuong Anh Bui

Example of implemented green facade in Milan

References:

1. Mahmoud, I. and Morello, E. (2018) Co-Creation Pathway as a catalyst for implementing Nature-based Solution in Urban Regeneration Strategies Learning from CLEVER Cities framework and Milano as test-bed.

2. Mahmoud, I. and Morello, E. (2020) Are Nature-based solutions the answer to urban sustainability dilemma? The case of CLEVER Cities CALs within the Milanese urban context.

3. Milano Clever Cities (2019) La Fermata Tibaldi, nuova centralità verde, CLEVER Milano. Available at: https://milanoclever.net/cal-3/ (Accessed: 1 March 2024)

Photo by Chiara Parretta.

EU_11 Opération-Récréation: Schoolyard Greening in Brussels

Context Overview

NBS TYPOLOGY Restoration	ECOSYSTEM TYPE Green spaces (gen.); Agriculture & food spaces
IMPACT SCALE Site	URBAN TYPOLOGY Urban core
SOCIAL CHALLENGES ADDRESSED Flooding; Biodiversity loss; Social wellbeing; Nature connection	CO-BENEFITS Place regeneration & urban revitalizations

The Brussels-Capital Region (BCR) is a densely populated urban area with 7,441 inhabitants per square-kilometer, almost 20 times the national average (Brussels Environment, 2022). Although 52 percent of its surface is covered by vegetation, public green spaces and accessible vegetated areas only cover 19.7 percent of the BCR's territory, benefiting mainly the well-off south and southeast neighbourhoods of the Region (Brussels Environment, 2023). The greening agenda of the city relies on two main policies, the Green Corridor (maillage vert), initiated in 1997 and aimed at maintaining and restoring connectivity between natural habitats throughout the territory (Brussels Environment, 2023), as well as the Regional Nature Plan aimed at establishing specific nature conservation goals and urban greening strategies by 2050 (Brussels Environment, 2016).

Introduction to the case study

Opération-Récréation aims at greening school playgrounds to meet biodiversity and rainwater storage needs in Brussels. The Opération-Récréation project was launched in 2021 by Brussels Environment – the regional administration for Environment and Energy (Brussels Environment, 2021) as a pilot for nature-based schoolyard transformation in the BCR. It addresses critical challenges of the BCR encompassing biodiversity loss, urban heat island effect, heightened flood risks, and socio-economic disparities. In this context, the Opération-Récréation pilot project embraces a nature-based solutions approach, notably through the conceptualization and implementation of Nature-based Climate School Shelters (NbCSS) (Ruiz-Mallén et al., 2023).

Nature-based Solutions characteristics

Opération-Récréation's main goal is to transform the schoolyards of 19 Brussels schools into high-quality, mixed-use, green play and resource areas while bringing refreshing nature into the neighbourhood (Brussels Environment, 2021). This transformation involves the removal of impermeable surfaces and the introduction of diverse green elements, such as climbing trees, fruit trees, bird and insect houses, wildflowers, vegetable gardens, small water bodies, fountains, and wooden play structures. This project also aims at rendering green spaces more accessible to disadvantaged families through the opening of the schoolyards to the public (notably during heatwaves), acting thereby as cooling islands, beneficial for the whole neighbourhood (Brussels Environment, 2021).

Governance

The implementation of NbCSS is achieved through a participatory process with diverse stakeholders within the educational ecosystem, from educators and administrators to parents' associations and schoolchildren. The participatory approach ensures that the transformation aligns with the unique characteristics and needs of each schoolyard. The comprehensive methodological and financial support provided, totalling circa 300,000 EUR (354,000 USD) per school, underscores the commitment to effective governance (Brussels Environment, 2021). The timeline, spanning from school selection in July 2021 to the anticipated completion by the end of 2024, adheres to a meticulous series of stages, emphasizing diagnostics, awareness, planning, and execution.

Outcome

Four green schoolyards have already been completed as part of the Opération-Récréation project (Brussels Environment, 2025) (Fig 1). The project's strengths lie in its interdisciplinary nature. Challenges, such as escalating costs and timeline considerations, underscore the complexities inherent in current large-scale urban interventions. Regular updates on the project's progress are accessible online (Bubble.Brussels, 2024), maintaining transparency and engagement. In essence, Opération-Récréation embodies a comprehensive and forward-thinking endeavour that showcases the potential of nature-based solutions in addressing urban climate-related challenges while concurrently fostering environmental education and children's wellbeing through increased contact with nature (Fig.2).

Writing: Elsa Gallez
Review: Veronica Olivotto
Copy Editing: Veronica Olivotto
Design: Phuong Anh Bui

Biodiversity audit in the schoolyard – Schoolchildren identifying places that they like and dislike, at Ecole en couleurs, Brussels (BE). Bénédicte Maindiau for Brussels Environment. https://www.bubble.brussels/ecole-encouleurs/

References:

1. Brussels Environment (2016) Plan régional nature 2016-2020 en Région de Bruxelles-Capitale. PLAN NATURE. Available at: https://environnement.brussels/sites/default/files/user_files/prog_20160414_naplan_fr.pdf (Accessed: 29 July 2025).
2. Brussels Environment (2021) Opération Ré-Création: Appel à Projets pour la Végétalisation des Cours D'école en Région Bruxelloise.
3. Brussels Environment (2022) État de la Nature en Région de Bruxelles-Capitale Synthèse 2022. Available at: https://document.environnement.brussels/opac_css/elecfile/SYNTH_Etat_Nature_2022_FR_DEF.pdf (Accessed: 29 July 2025).
4. Brussels Environment (2023) Espaces verts et leur gestion. Available at: https://environnement.brussels/citoyen/outils-et-donnees/etat-des-lieux-de-lenvironnement/espaces-verts-et-leur-gestion#espaces-verts-accessibilite-au-public (Accessed: 29 July 2025).
5. Brussels Environment (2025) Opération Ré-création : 4 chantiers terminés. Available at: https://environnement.brussels/enseignement/news/2025/operation-re-creation-4-chantiers-termines (Accessed: 29 July 2025).
6. Bubble.Brussels (2024) Opération Ré-création » Bubble, Bubble.brussels. Available at: https://www.bubble.brussels/operation-re-creation/ (Accessed: 17 December 2024).
7. Ruiz-Mallén, I. et al. (2023) 'Nature-Based Solutions for Climate Adaptation in School Environments: An Interdisciplinary Assessment Framework', in Z. Allam (ed.) Sustainable Urban Transitions: Research, Policy and Practice. Singapore: Springer Nature, pp. 87–105. Available at: https://doi.org/10.1007/978-981-99-2695-4_6.

EU_12

ClearingHouse: Post-Industrial Urban Forests in Gelsenkirchen

Context Overview

NBS TYPOLOGY	ECOSYSTEM TYPE
Restoration	Forests & trees
IMPACT SCALE	**URBAN TYPOLOGY**
Greater urban region	Peri-urban
SOCIAL CHALLENGES ADDRESSED	**CO-BENEFITS**
Nature connection; Physical and mental health	Place regeneration & urban revitalization; Community engagement & social cohesion; Reinforcement of socio-ecological knowledge

The ClearingHouse project applies a Living Lab approach to promote urban forestry in two industrial landscapes of the Rhine-Ruhr region. Gelsenkirchen is part of this polycentric metropolitan area and home to over 265,000 people, was once dubbed the "City of a Thousand Fires" for its coal and steel industries. As in many post-industrial cities, demand for accessible green spaces has grown—especially following the pandemic. Decades of social and economic change have left large areas of derelict land that have naturally regenerated into forests. In Gelsenkirchen, around 20 percent of forested areas are on former industrial sites and slag heaps (Clearing House, 2020). These novel ecosystems, strategically located within the urban core, represent valuable assets for addressing social and ecological challenges.

Introduction to the case study

The HORIZON 2020-funded ClearingHouse project (Collaborative Learning in Research, Information-sharing and Governance on How Urban tree-based solutions support Sino-European urban futures) in Gelsenkirchen aimed to gain insights into the social and ecological performance of post-industrial urban green spaces that have undergone natural succession to forests. The goal was to leverage the potential of post-industrial natural succession as urban forest nature-based solutions—tree-based urban ecosystems that address societal challenges whilst simultaneously providing ecosystem services for human well-being and biodiversity benefits (Dumitru et al., 2021).

Nature-based Solutions characteristics

The nature-based solutions examined in the Gelsenkirchen case study of the Clearing House project focus on green inner-city redevelopment through natural succession on post-mining land (Roitsch et al., 2024). Specifically, the project highlights two sites: the Rheinelbe industrial forest and the former coal mine Zeche Hugo, both located near Gelsenkirchen's city center (Clearing House, 2023). These sites exemplify post-industrial landscapes at different stages of natural succession, with varying canopy cover densities. The Rheinelbe and Zeche Hugo sites illustrate this successional gradient, ranging from early-stage areas dominated by Betula pendula and Crataegus monogyna to more mature ecosystems characterized by Fagus sylvatica and Quercus robur (Kraus et al., 2022).

These two post-industrial urban green spaces are part of the Emscher Landscape Park (ELP), a broader, regional cooperation project in the northern Ruhr area to create the world's largest connected park system (450 square-kilometers) (Interlace Hub, 2023). Using a combination of social and ecological methods, these two living labs will explore how Urban Forestry and Nature-based Solutions (UF-NbS) can be employed to restore former mining sites, promote urban regeneration, and support sociocultural integration.

Governance

Gelsenkirchen's strategy has focused on prevention, education and participation as well as the idea of "learning city"—the combination of integrative, equal opportunity and good education with "Education for Sustainable Development (ESD)". In line with this, an important part of the ClearingHouse project was the development and co-creation workshops guidelines for all frontrunner cities, including Gelsenkirchen. The aim of these co-creation workshops was to meaningfully engage with people in the city, local experts, practitioners, and decision makers to bring all the local knowledge and perceptions together on how to best improve and regenerate the state of urban forests to benefit people and nature. More specifically, in Gelsenkirchen, the aim was to elicit the differing demands, perceptions and opinions of a highly diverse group of people to propose strategic targets for post-industrial forest that are not only ecologically sound but also socially acceptable for the local community (Clearing House H2020, 2022). As part of the public participatory engagement, participatory mapping methods were also used to empower local residents to spatially identify positive and negative perceptions in relation to post-industrial urban forests so as to inform planning and management decision-making (Clearing House 2023). In addition, in order to learn more about the ecological performance and potential of secondary, post-industrial forests for biodiversity conservation, two Marteloscopes were installed in the Rheinelbe industrial forest and the former coal mine of Zeche Hugo to monitor the development and delivery of ecosystem services of these sites over time.

The share of funding of the Clearing House project for the Gelsenkirchen corresponded to a total of roughly 923,000 EUR (1 million USD), of which 758,437 EUR (984,955 USD) went to the research partner (Humboldt-Universität zu Berlin) and 160,968 EUR (189,942 USD) for The City of Gelsenkirchen (CORDIS 2024). The International Building Exhibition Emscher Park (IBA), with support from NRW Urban GmbH & Co. KG initiated the project by providing land, buildings, and administrative help. The Ministries of Environmental Protection and Urban Development endorsed the idea with financial backing. The Forest Authority and Regional Forest Office Ruhrgebiet employed three foresters for area maintenance, while research was managed by the Biological Station Western Ruhr Area and the Humboldt-Universität zu Berlin (Clearing House, 2024).

Outcome

Using a combination of social and ecological methods, these two living labs will explore how Urban Forestry and Nature-based Solutions (UF-NbS) can be employed to restore former mining sites, promote urban regeneration, and support sociocultural integration. An important part of the ClearHouse project is the development and co-creation workshops guidelines for all frontrunner cities, including Gelsenkirchen. More specifically, in Gelsenkirchen, the aim was to elicit the differing demands, perceptions and opinions of a highly diverse group of people to propose strategic targets for post-industrial forest that are not only ecologically sound but also socially acceptable for the local community (Clearing House H2020, 2022). As part of the public participatory engagement, participatory mapping methods were also used to empower local residents to spatially identify positive and negative perceptions in relation to post-industrial urban forests so as to inform planning and management decision-making (Clearing House, 2023).

One of the main challenges identified in this project is the budgetary cuts and lack of economic resources from the local administration, which can severely hinder the establishment and future planning and maintenance of urban forests in the studied region. In addition, one of the main challenges raised by local practitioners is the diversity of needs and expectations that different groups place on urban green spaces and the lack of frameworks or tools, know-how to manage and balance these. Similarly, the growing population in these post-industrial areas and the increasing demand for housing have intensified the pressure on urban land. This makes it difficult to prioritize and protect post-industrial sites for urban forest nature-based solutions.

Writing: Oriol Garcia-Antúnez
Review: Veronica Olivotto
Copy Editing: Veronica Olivotto
Design: Phuong Anh Bui

View of the wastelands undergoing natural succession in Gelsenkirchen

References:

1. Clearing House (2024) Case study info sheets, CLEARINGHOUSE. Available at: https://clearinghouseproject.eu/ (Accessed: 1 July 2024).
2. Clearing house (2023) Blog—Insights from the CLEARING HOUSE Co-Design Workshop in Gelsenkirchen by Ian Whitehead, European Forest Institute (EFI) Accessed on July 2024 on https://clearinghouseproject.eu/2023/09/; site inactive on 28 April 2025.
3. CORDIS (no date) 'CLEARING HOUSE—Collaborative Learning in Research, Information-sharing and Governance on How Urban tree-based solutions support Sino-European urban futures'. Available at: https://cordis.europa.eu/project/id/821242 (Accessed: 17 December 2024).
4. Dumitru, A. et al. (2021) Evaluating the Impact of Nature-based Solutions: A Handbook for Practitioners. Available at: https://doi.org/10.13140/RG.2.2.10757.47843.
5. How can trees play a role in cities' social changes? Gelsenkirchen Industrial Sites—CLEARING HOUSE (2022). Available at: https://www.youtube.com/watch?v=s1FW5l_ET1o (Accessed: 17 December 2024).
6. Interlace Hub (2022) Emscher Landscape Park Master Plan—Ruhrgebiet, Interlace Hub. Available at: https://interlace-hub.com/emscher-landscape-park-master-plan-ruhrgebiet (Accessed: 17 December 2024).
7. Kraus, D. et al. (2022) 'Spatially explicit database of tree related microhabitats (TreMs)'. Institut national de recherche pour l'agriculture, l'alimentation et l'environnement (INRAE). Available at: https://doi.org/10.15468/OCOF3V.
8. Roitsch, D. et al. (2024) 'Co-production of urban forests as nature-based solutions: Motivations and lessons-learnt from public officials', Environmental Science & Policy, 157, p. 103764. Available at: https://doi.org/10.1016/j.envsci.2024.103764.

Photo by Ian Whitehead.

EU_13

The BiodiverCity Project: Biodiversity Conservation in Malmö

Context Overview

LOCATION

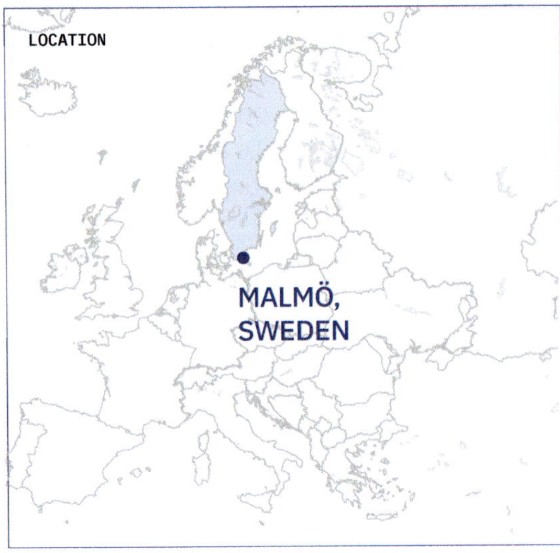

NBS TYPOLOGY	ECOSYSTEM TYPE
Restoration; Constructed & novel ecosystems	Green spaces (gen.); Stormwater infrastructure; Roofs & walls; Riparian areas
IMPACT SCALE	**URBAN TYPOLOGY**
City	Urban core
SOCIAL CHALLENGES ADDRESSED	**CO-BENEFITS**
Biodiversity loss; Social inclusion	Flood mitigation; Human health and well-being; Urban cooling; Reduced infrastructure cost; Air quality

Malmö is the third largest city in Sweden, and the largest city in the Swedish county of Scania. The Municipality of Malmö has around 350,000 citizens and its Metropolitan Region around 700,000. The city is characterized by an oceanic climate, with mild summers (average temperatures between 12 and 22°C), occasional heat waves, and fairly cold and windy winters (average temperatures between −3 and 4°C). Rainfall is light to moderate throughout the year (average of 169 wet days). Malmö is one of most-industrialized cities in Scandinavia and currently Sweden's fastest growing city.

Introduction to the case study

BiodiverCity worked at the city scale aiming to create new green areas to promote biodiversity conservation, and rehabilitate/restore species and degraded ecosystems within Malmö. The project BiodiverCity intended to address several key challenges that commonly affect urban areas, particularly in densely urbanized cities. These challenges included the degradation of natural/semi-natural ecosystems, lack of green spaces, and lack of opportunities to boost urban biodiversity. To do so, BiodiverCity developed products, services, and processes for a greener, healthier, and aesthetically attractive city. The Sustainable Development Goals addressed in this project include SDG15—Green space, habitats, and biodiversity, and SDG3—Improving physical health.

Nature-based Solutions characteristics

As part of the activities related to the conservation of biodiversity, this project created new habitats and avoided ecosystem damage, took specific measures to protect species, and developed capacity building, as a means for conservation governance. The nature-based solutions created/restored in this study included green roofs, green walls/facades, alley and green streets, community gardens and allotments, green areas for water management, and sustainable urban drainage systems. Specific examples of implementation activities included the creation of green roofs on the building of Malmö Service Administration and green walls on a commercial building on Västra Hamnen.

Governance

The Municipality of Malmö initiated the organization of BiodiverCity as a response to European and city environmental policies/strategies. The management set-up included co-governance with government and non-government actors. The sources of funding for this project included the European Union and public national budgets, having a total cost from 500,000 EUR (645,000 USD) to 2 million EUR (2.5 million USD) The main beneficiaries included several actors, such as the local government, the private sector, researchers, citizens, and community groups. During the six years of the project (2012-2018) a wide range of stakeholders from many sectors, including ecologists, landscape architects, scientists, entrepreneurs, and developers were brought together to work on specific greenery projects. In addition, the local community was involved through participatory events that included co-planning, consultation (for example, workshop, surveys), and joint implementation (for example, tree planting). As a result of this multidisciplinarity, the created habitats were able to meet both ecological requirements and requirements related to operation and maintenance, installation costs, and aesthetics.

Outcome

BiodiverCity developed products, services, and processes to support and enhance the city's biodiversity, including innovative ways to design greenery in street trees, green roofs, façades and walls, mobile plant systems, urban biotopes, and three-dimensional greenery. The specific co-benefits of this project can be categorized into environmental impacts, socio-cultural impacts, economic impacts, and health and wellbeing. The environmental impacts include promoting naturalistic styles of landscape design for urban development, increased green area and species diversity, and the delivery of important ecosystem services such as stormwater management and heat mitigation. The socio-cultural impacts are related to social justice and cohesion due to improved access to green spaces and healthy food. The economic impacts include the commercialization of green solutions and reduced maintenance costs for households where greenery was installed. The health and wellbeing impacts include the creation of opportunities for recreation and exercise activities. Another useful outcome of this project is the enhanced knowledge of the more suitable plants and solutions in specific framework conditions. Some challenges that this project faced were related to a lack of knowledge/experience with living walls (from the public, potential new customers, and industry), which can be detrimental to the development of living walls in Sweden, and the need for specialist knowledge to address complex irrigation systems.

Writing: Filipa Grilo
Review: Veronica Olivotto
Copy Editing: Veronica Olivotto
Design: Phuong Anh Bui

References:

1. Smart City Sweden (no date) BioDiverCity – Green Areas to Boost Sustainability, Smart City Sweden. Available at: https://smartcitysweden.com/best-practice/342/biodivercity-green-innovation-in-urban-environments/ (Accessed: 25 January 2024).

2. Urban Nature Atlas (2021) BiodiverCity, Urban Nature Atlas. Available at: https://una.city/nbs/malmo/biodivercity (Accessed: 25 January 2024).

Photo by Oriol Garcia-Antúnez.

EU_14 Connecting Nature: An Urban Gardens Green Network

Context Overview

LOCATION

NBS TYPOLOGY	ECOSYSTEM TYPE
Constructed & novel ecosystems	Agriculture & food spaces
IMPACT SCALE	**URBAN TYPOLOGY**
City	Urban core
SOCIAL CHALLENGES ADDRESSED	**CO-BENEFITS**
Social wellbeing; Nature connection	Community engagement & social cohesion; Reinforcement of socio-ecological knowledge

The *ecoHortas* are multifunctional urban gardens that improve environmental quality, foster social participation, and enhance citizens' well-being in A Coruña. Situated on Spain's northwest coast, A Coruña is a coastal metropolis in the Ártabro Gulf. This compact city, spanning less than 40 square kilometers, is home to approximately 244,000 residents. Boasting a rich architectural legacy, A Coruña is complemented by its natural and scenic landscapes. The city's economic foundation primarily rests on its service industry, which encompasses sectors like tourism and maritime commerce. Despite its economic and historical significance, A Coruña faces common urban challenges, notably a shortage of green space within its limited geography, compounded by low biodiversity, heat stress, sparse vegetation, and urban waste management pressures.

Introduction to the case study

A Coruña was selected as one of the cities representing the Connecting Nature project due to its need to create more green spaces and its vision for a gentler mobility system prioritizing green areas. This initiative aimed not only to develop green infrastructure compatible with cultural heritage but also to significantly enhance resident well-being and tourism appeal. Participation in projects like Connecting Nature provided A Coruña with valuable insights for advancing towards an integrated lifestyle based on ecosystems, sustainable urban planning, and health. While the nature-based solutions concept is relatively new to A Coruña, the city has some precedents from the 1990s and 2000s. Notable examples include the transformation of the city's main landfill into one of its largest parks and the conversion of former military lands into a coastal green area. This approach represents a shift from the traditional concept of parks and green spaces, demonstrating A Coruña's previous commitment to sustainable urban development that integrates nature into the city fabric, culminating with the adoption of a preliminary Green Infrastructure Plan in 2018.

Nature-based Solutions characteristics

This case study focuses on the ecoHortas Urban Gardens program, launched in 2018 as a key component of the shift toward nature-based solutions. The program established three main gardens—Eirís, Agra do Orzán, and Novo Mesoiro—offering 218 plots for citizen use. Although not originally designed using the Connecting Nature framework, which did not exist at the time, the gardens were conceived as nature-based solutions delivering multi-level benefits. As a fast-follower city in the project, A Coruña later identified areas for improvement, including the governance model and the absence of common social spaces.

Governance

ecoHortas are established on municipal land—abandoned plots or underused park areas—and their locations were determined through citizen participation. Plots are assigned to citizens and associations for two to three years, with some reserved for educational centers, non-profit groups, retirees, low-income households, and people with special needs. The municipality provides water and shared tools, and users must practice organic agriculture. To support this initiative, A Coruña facilitated self-management of the gardens through workshops and training sessions, helping to establish operational norms and management committees.

Funding for ecoHortas comes from various sources. The city relies heavily on its budget for initial investments and ongoing operational costs, with additional financial backing from European, national, or regional sources. Locally, the city allocates municipal funds for parks and gardens and has implemented participatory budgets of 2 million EUR annually. The Connecting Nature project itself had a total cost of 11.7 million EUR (13.8 million USD), with NET EU contributions of EUR 531.5 million EUR (627 million EUR) for the University of A Coruña and 334.6 million EUR (394 million USD) for the Municipality of A Coruña as primary beneficiaries of the case study. At a policy level, A Coruña has crafted an integrated strategy for sustainable urban development (EIDUS) supported by European Regional Development Fund (ERDF) resources. The city is also among the first in Spain to implement the new Spanish Urban Agenda locally, aligning with the European Urban Agenda and the UN Sustainable Development Goals.

Outcome

A Coruña's urban gardening and green space initiative represent a significant step towards creating a more sustainable and resilient city. The ecoHortas program and the broader shift towards nature-based solutions demonstrate the city's commitment to environmental stewardship, social cohesion, and citizen engagement. These efforts have transformed underutilized spaces into community assets and contributed to biodiversity, climate change mitigation, and improved quality of life for residents.

While A Coruña's efforts in urban gardening and green space initiatives are highlighted, they also underscore the complexities of implementing large-scale environmental changes in urban settings. The city's hierarchical structure and lack of inter-departmental coordination often result in siloed approaches, potentially leading to duplicated efforts or conflicting initiatives. This structural limitation highlights the need for a more integrated urban planning and management approach. The urban gardens' management model also faces challenges, with the need to balance citizen involvement and municipal oversight. The city's revised garden allocation process, which encourages gardeners to self-organize while maintaining some municipal control, represents an innovative approach to addressing this issue. However, building trust and capacity within the gardeners' association remains an ongoing process.

References:

1. Collier, M.J. et al. (2023) 'An integrated process for planning, delivery, and stewardship of urban nature-based solutions: The Connecting Nature Framework', Nature-Based Solutions, 3, p. 100060. Available at: https://doi.org/10.1016/j.nbsj.2023.100060.

2. Connecting Nature (2022) Deliverable 14: Report on implementation of Connecting Nature Frameworks in the Fast Follower Cities.

3. Connecting Nature (2023) A Coruña: An Urban Gardens Green Network.

Writing: Jarumi Kato Huerta
Review: Veronica Olivotto
Copy Editing: Veronica Olivotto
Design: Phuong Anh Bui

EU_15 URBiNAT: Sofia's "Healthy Corridor"

Context Overview

NBS TYPOLOGY	ECOSYSTEM TYPE
Restoration; Constructed & novel ecosystems	Corridors; Green spaces (gen.)
IMPACT SCALE	**URBAN TYPOLOGY**
Neighbourhood/Community	Urban core
SOCIAL CHALLENGES ADDRESSED	**CO-BENEFITS**
Nature connection; Physical and mental health	Place regeneration and urban revitalization

Sofia, the capital of Bulgaria, experienced rapid industrialization and urban growth after 1960, leading to the construction of around 15 large housing estates (LHEs), mostly on former farmland at the city's edge. Today, these estates house about 575,000 people—nearly half of Sofia's population—and account for 47.3 percent of the city's housing stock (Tasheva-Petrova et al., 2021). The post-1990 transition to a market economy exposed major challenges in these areas, including shifts in property ownership, reduced public maintenance, and growing social inequalities. Public spaces such as sidewalks, pedestrian zones, and green areas have come under strain. While trees planted during the original development still offer environmental benefits, many other green features have deteriorated. Despite oversight by Sofia's "Green System" Department, budget constraints have led to uneven maintenance and limited access to quality green spaces in LHEs.

Introduction to the case study

URBiNAT uses a Living Lab approach to build a linear park intervention bringing social, mental health and ecological benefits. The URBiNAT project established Living Labs in Porto, Nantes, and Sofia to promote the "Healthy Corridor" concept, using nature-based solutions to transform public spaces, foster community, and improve physical and mental health. In Sofia, the Living Lab was located in the Nadezhda district, in the north-central part of the city. This area spans 115.16 hectares and consists of five neighborhoods, where 37,770 residents live in 17,069 dwellings.

Nature-based Solutions characteristics

The Nadezhda Living Lab became a key component of the URBiNAT project, serving as a testing ground for implementing nature-based solutions and community-led planning to improve urban life. Public spaces in Nadezhda vary in size and function but have faced safety issues and diminishing availability. Nonetheless, comprehensive urban plans adopted between 2013 and 2015 helped safeguard undeveloped public spaces. The Healthy Corridor was envisioned as a two-point-five kilometer linear park that connects Severen Park with Park Nadezhda. Along its route, four distinct intervention areas were created: 'Co-Place', 'Health Energy', 'Aqua Vita', and 'Green Assembly'. Additional paths branch off the main route to expand access and integrate the park into the broader urban fabric. This strategy combines different spatial and community needs, extending the benefits of the project beyond the immediate neighborhood and creating a more inclusive urban experience. The corridor emphasised the importance of mid-scale public spaces in meeting collective needs. Alongside enhancing physical infrastructure, the project sought to tackle social and economic inequalities, promote mental and physical health, improve daily living conditions, and help regulate the local microclimate.

Governance

The URBiNAT project in Sofia had a total budget of 13.6 million EUR (16.1 million USD), with 13 million EUR (15.3 million USD) provided by the European Union. The project ran from June 1, 2018, to March 13, 2024. Sofia's implementation of URBiNAT combined traditional municipal management with innovative participatory approaches. The city government led the initiative, working across several levels of local administration. The project also partnered with the University of Architecture, Civil Engineering and Geodesy, incorporating academic knowledge into urban planning and design. Community involvement was central to the project. The Nadezhda Living Lab encouraged collaboration between city authorities and residents in shaping the Healthy Corridor. However, balancing this participatory model with existing top-down administrative processes proved challenging. There were occasional tensions between adhering to official procedures and the flexibility needed for inclusive, responsive engagement. Still, the city's project team demonstrated adaptability and persistence, continually refining their approach to meet emerging needs. However, coordination remained a challenge—both within the team and in communication with external stakeholders. This governance model reflects the complexity of combining established bureaucratic systems with more dynamic, community-driven processes. The ongoing challenge lies in aligning institutional structures with the goal of creating more responsive, people-focused urban development.

Outcome

The URBiNAT initiative in Sofia achieved notable results by applying the Healthy Corridor concept. A total of 937 participants, both adults and youth, engaged in workshops and co-design sessions. These participatory activities ensured that community needs were reflected in the project's design and implementation (Nunes et al., 2021). The project team organized suggestions into specific action categories that involved all stakeholders. A comprehensive strategy for the Healthy Corridor was developed in alignment with Sofia's broader urban planning goals. Out of 43 proposals, 23 nature-based solutions were integrated into the final plan, 11 were adopted by other municipal projects, and 10 were not implemented. This means that 52 percent of all suggested solutions were fully realized, with another 25 percent having the potential to be developed in future initiatives, highlighting the project's ongoing relevance and scalability. A preliminary urban development plan was created, organizing interventions across 16 subzones grouped into four major zones and two smaller zones. This plan encompassed 138.1 hectares across four housing estates. The city's Municipal Council formally approved the project. An Advisory Board was established to guide the inclusive implementation and ensure the long-term management of the corridor. The project also fostered strong working relationships among various municipal departments, leading to the institutionalisation of the Healthy Corridor concept in a purposefully created municipal unit.

Writing: Jarumi Kato Huerta
Review: Veronica Olivotto
Copy Editing: Veronica Olivotto
Design: Phuong Anh Bui

References:

1. Nunes, N., Björner, E. and Hilding-Hamann, K.E. (2021) 'Guidelines for Citizen Engagement and the Co-Creation of Nature-Based Solutions: Living Knowledge in the URBiNAT Project', Sustainability, 13(23), p. 13378. Available at: https://doi.org/10.3390/su132313378.

2. Tasheva-Petrova, M. et al. (no date) 'Re-claiming space for public life: messages from the northwestern periphery of Sofia', Urbani Izziv, 32, pp. 91–105. Available at: https://doi.org/10.5379/urbani-izziv-en-2021-32-supplement-6.

3. URBiNAT (no date) София, България, URBiNAT. Available at: https://urbinat.eu/cities/sofia/ (Accessed: 1 March 2024).

EU_16

Flower Power: A Community-run Wildflower Nursery in Glasgow

Context Overview

NBS TYPOLOGY Restoration	ECOSYSTEM TYPE Green spaces (gen.)
IMPACT SCALE Neighbourhood/Community	URBAN TYPOLOGY Urban core
SOCIAL CHALLENGES ADDRESSED Social inclusion; Biodiversity loss; Heat	CO-BENEFITS Place regeneration and urban revitalization

Flower Power is an ecological restoration project tackling urban meadows in marginalized areas of Glasgow. Glasgow, located in west-central Scotland, is the country's second-largest city with a population of over 600,000 (UK Population Data, 2023). The city's industrial past and subsequent population decline have contributed to various long-term challenges, including social inequality and environmental degradation, issues that are increasingly being intensified by the impacts of climate change. Many of Glasgow's residents, particularly those in disadvantaged communities, face barriers related to poor health, limited access to quality green spaces, and social isolation. At the same time, environmental threats such as biodiversity loss, heatwaves, flash flooding, and drought are becoming more frequent, affecting the city's most vulnerable populations.

Introduction to the case study

Glasgow City Council has adopted a nature-based solutions approach to tackle these complex issues. This strategy aims to address health, environmental, and social problems, especially those impacting marginalized communities, by implementing small and medium-scale community garden projects alongside integrated Sustainable Urban Drainage Systems (SUDS). One example is 'Flower Power', a locally led nature-based solutions initiative in Pollok Country Park, which is Glasgow's largest and only Country Park, encompassing 146 hectares. The park was gifted to the city in 1966 and is managed by Glasgow City Council, making it public land. The wildflowers grown in the nursery are planted in various public spaces throughout Glasgow, further emphasising the project's role in enriching publicly accessible green areas. Developed under the EU Horizon 2020 Connecting Nature Project, this initiative aims to restore meadows and support pollinators, such as bees and butterflies. Simultaneously, it fosters a deeper connection between people and nature, providing various social benefits. These benefits include enhanced well-being and public health, increased social cohesion, income-generating opportunities, and greater public participation in local planning and decision-making.

Nature-based Solutions characteristics

The community-run wildflower nursery at Pollok Country Park cultivates approximately 10,000 native wildflowers each year, which are then planted in various public parks and green spaces across the city. This effort aims to reverse the decline of meadow and insect species by reintroducing native flora that provides essential habitats and foraging resources for pollinators, such as bees and butterflies. Socially, the initiative engages local communities in hands-on environmental stewardship, offers skills training, and fosters inclusion, particularly among vulnerable groups. Economically, the project contributes to cost-effective green infrastructure by reducing procurement expenses for native plants and creating opportunities for local employment and volunteering.

Governance

The Flower Power project is overseen by Glasgow City Council, with strong support from volunteers and partner organizations such as The Conservation Volunteers Scotland (TCV Scotland) and Grow Wild. It was part of the EU Connecting Nature Horizon 2020 Research and Innovation Action, a programme designed to help cities deliver effective nature-based solutions that address their most urgent challenges. The overall Connecting Nature project had a budget of 11.7 million EUR (13.2 million USD), with Glasgow receiving 748,250 EUR (845,522 USD) as the lead local partner for this case study. While the programme ran from June 1, 2017, to May 31, 2022, the vision for these nature-based solutions projects was designed to extend over 15 years. The long-term goals include generating evidence-based outcomes, encouraging entrepreneurship, particularly among young people, and strengthening the city's appreciation and stewardship of its natural environment.

Flower Power and other related projects across the city were integrated into the Glasgow Open Space Strategy, which is grounded in detailed local data and spatial analysis. This strategy helps identify open spaces, assess their quality, and meet the specific needs of local communities. It promotes the use of nature-based design and management approaches to enhance these spaces through innovation and community engagement. The Council employed a Local Context Analysis method to translate this strategy into practice, focusing on 15 areas within Glasgow. This allowed planners to tailor projects to the unique social and spatial characteristics of each neighborhood. The approach also emphasized community participation, encouraging residents to play an active role in shaping and maintaining green spaces, which helps build a sense of ownership and collaboration.

Outcome

One of the primary challenges faced by the Flower Power initiative and the broader Connecting Nature programme was how to manage public green spaces during a period of public sector austerity in the UK. There was a need to maintain the quality and accessibility of public spaces while also finding more sustainable, cost-effective ways to implement nature-based solutions, and involve communities in their upkeep. Despite these challenges, Flower Power has had a lasting impact. It continues to operate a year-round volunteer programme aimed at fostering social cohesion and community participation. Each year, the project cultivates 10,000 native wildflowers, which are used in parks, green spaces, and community projects throughout the city. The nursery and garden established through the initiative have become a vital source of native plant species for biodiversity efforts across Glasgow. The benefits of the project are likely to persist due to the Glasgow Pollinator Plan, which aligns with the Scottish National Pollinator Strategy and supports ongoing conservation efforts in the region.

Another key outcome has been in environmental education. A public awareness campaign was launched to emphasize the value of wildflower meadows and to help residents understand why some areas that were previously maintained as closely mown lawns are now allowed to grow more naturally. These educational efforts are helping to shift public attitudes and foster greater community support for biodiversity-friendly landscapes.

Writing: Jarumi Kato Huerta
Review: Veronica Olivotto
Copy Editing: Veronica Olivotto
Design: Phuong Anh Bui

References:

1. Climate Ready Clyde (2021) Glasgow City Region Climate Adaptation Strategy and Action Plan.

2. Collier, M.J. et al. (2023) "An integrated process for planning, delivery, and stewardship of urban nature-based solutions: The Connecting Nature Framework," Nature-Based Solutions, 3, p. 100060. Available at: https://doi.org/10.1016/j.nbsj.2023.100060.

3. Frantzeskaki, N. et al. (2020) "Examining the policy needs for implementing nature-based solutions in cities: Findings from city-wide transdisciplinary experiences in Glasgow (UK), Genk (Belgium) and Poznań (Poland)," Land Use Policy, 96, p. 104688. Available at: https://doi.org/10.1016/j.landusepol.2020.104688.

4. Glasgow case studies (no date) Connecting Nature. Available at: https://connectingnature.eu/city-case-studies/glasgow(Accessed: December 17, 2024).

5. Glasgow's Flower Power (no date) Glasgow City Council. Available at: https://www.glasgow.gov.uk/article/4893/Glasgow-s-Flower-Power (Accessed: December 17, 2024).

6. Oppla, J. (2019) Pollok Park Flower Power to the People, Glasgow—social cohesion wildflower nursery, Connecting Nature. Available at: https://connectingnature.oppla.eu/casestudy/19383 (Accessed: December 17, 2024).

7. O'Sullivan, F., Mell, I. and Clement, S. (2020) "Novel Solutions or Rebranded Approaches: Evaluating the Use of Nature-Based Solutions (NBS) in Europe," Frontiers in Sustainable Cities, 2. Available at: https://doi.org/10.3389/frsc.2020.572527.

EU_17 Green Krupnicza: Blue-Green Infrastructure in Kraków

Context Overview

LOCATION

NBS TYPOLOGY	ECOSYSTEM TYPE
Constructed & novel ecosystems	Green spaces (gen.); Stormwater infrastructure; Forests & trees; Marine (gen.)
IMPACT SCALE	**URBAN TYPOLOGY**
Street	Urban core
SOCIAL CHALLENGES ADDRESSED	**CO-BENEFITS**
Heat	Flood mitigation

Located by the Vistula River in southeastern Poland, Kraków is home to approximately 804,000 (Statistics Poland, 2023). It experiences a humid continental climate with cold winters, warm summers, and moderate precipitation throughout the year. Kraków is notable for its rich historical heritage, including a well-preserved medieval core, and its role as a cultural and academic hub. The city's urban landscape features a combination of historical architecture, post-industrial zones, and a growing emphasis on green infrastructure as part of wider efforts to address environmental degradation and enhance urban liveability.

Introduction to the case study

The "Green Krupnicza" project in Kraków, officially opened in 2022, exemplifies a pioneering effort to implement blue-green infrastructure in a dense, historic urban setting. The aim of the project was to minimize the negative effects of climate change—long periods of drought and heat and to counteract the urban heat island. Located next to the Stare Miasto (Old Town) district, the 400-meter-long Krupnicza Street was transformed from a fully paved, car-oriented corridor into a pedestrian-friendly, pacified space with rain gardens, permeable surfaces, and increased vegetation. The area is characterized by a mix of residential buildings, cultural institutions, and commercial establishments. The initiative aimed not only to improve stormwater management and urban biodiversity but also to enhance social well-being and environmental resilience.

Nature-based Solutions characteristics

The "Green Krupnicza" project introduced blue and green infrastructure on a previously completely paved surface. It greened Krupnicza Street and turned it into a rain garden. As part of the investment, trees were planted along the entire length of the street. A total of 39 trees were planted of the "Plena" bird cherries (*Prunus avium var. Plena*) and black alders (*Sambucus nigra*). Approximately 6 m3 of structural substrate was prepared for each tree. Thanks to the use of structural soil, the tree root system has optimal conditions for development. Additionally, thousands of shrubs and perennials were planted on raised flowerbeds. Facade gardens were also designed and planted with hollyhocks (*Alcea spp.*). All the plants created a green mosaic. The investment also included the installation of tanks storing rainwater, which is used to water plants, as well as reducing the load on the city's sewage system.

Governance

The idea of the project was based on the concept of a 'woonerf street', i.e. a street with calm traffic, where the safety of pedestrians is a priority, while maintaining the possibility of cars moving. The introduction of a common space for all users, instead of dividing it into the road and sidewalks, and the presence of street furniture and greenery contributed to the creation of a friendly space for residents and visitors. Furthermore, information about the history of Krupnicza Street is emphasized on special boards placed along the street. The investment is the result of the cooperation between the municipal company managing green areas in Krakow (2.2 million EUR or 2.4 million USD), and numerous municipal entities, including infrastructure managers. The key partners of the project were primarily residents and non-governmental organizations. In preparation for the investment, numerous consultations, promotional campaigns and prototyping were carried out. Residents and entrepreneurs also became guardians of the gardens on the street. The transformation of the street was initiated by local activists. Their involvement included advocating for the street's renovation, organizing numerous community actions and initiatives involving residents. They also supported the promotion of the "Green Krupnicza" project, a continuation of the #3xTAKdlazieleni campaign in 2017 and beyond. The activists carried out outreach and educational activities, engaging the local community in the process of transforming the public space. Residents participated in consultations and continue to report maintenance issues on an ongoing basis. Local business owners on Krupnicza Street play a key role as plant caretakers, maintaining the façade gardens located next to their establishments.

Outcome

The reconstruction of Krupnicza Street was completed in May 2023. This historic street has completely changed its appearance, offering residents and visitors a new attractive green public space. Changing the existing paved surface to a rain garden created a green oasis, friendly to people and the environment. The garden slows down the outflow of rainwater into the sewage system, retains it in the soil, creates a favorable microclimate and counteracts the urban heat island. What is very important, residents actively participated in the transformation process.

Project Green Krupnicza on Kraków City Hall

Writing: Agnieszka Sobol
Review: Veronica Olivotto
Copy Editing: Veronica Olivotto
Design: Phuong Anh Bui

Acknowledgements:
Consultation Łukasz Pawlik

References:

1. Krakow.Pl (2023) 'Ulica Krupnicza oficjalnie otwarta!', 30 May. Available at: https://www.krakow.pl/aktualnosci/271491,26,komunikat,ulica_krupnicza_oficjalnie_otwarta_.html (Accessed: 17 December 2024).

2. Statistics Poland (2023) Bank Danych Lokalnych. Available at: https://bdl.stat.gov.pl/bdl/start (Accessed: 21 July 2025).

EU_18 Lisbon's Strategy for Urban Agriculture

Context Overview

LOCATION

NBS TYPOLOGY	ECOSYSTEM TYPE
Restoration	Agriculture & food spaces; Green spaces (gen.)
IMPACT SCALE	**URBAN TYPOLOGY**
Neighbourhood/Community	Urban core
SOCIAL CHALLENGES ADDRESSED	**CO-BENEFITS**
Food security; Social inclusion	Water quality; Recreational benefits

Lisbon, the capital of Portugal, is a coastal city on the north bank of the mouth of the Tagus River (38°43'00" N; 9°07'59" W). Lisbon has a Mediterranean climate, characterized by hot and dry summers and cold and rainy winters, with an annual mean temperature of 17.2 °C, mean annual precipitation of 704.8 mm (1960–2022 average; PORDATA 2023), and north, north-western prevailing winds. It covers an area of 8,545 ha, with a resident population of 545,000 people, surpassing one million people per day due to tourism and commuting movements.

Introduction to the case study

The development of a strategy for urban agriculture (2007) and the consequent creation of municipal horticultural parks intended to address the increasing number of dispersed and informal allotment gardens in Lisbon, especially after the economic crisis of 2008. Gardeners spontaneously create these informal spaces in municipal or private derelict land to unwind and connect with nature. These spaces, however, are not regulated nor supported by the municipality and are often in a precarious and unsanitary state. Therefore, by creating these parks, the municipality aims to respond to the need for urban agricultural spaces while providing better conditions to urban gardeners, increasing public and environmental health, and assuring greater food security to their users. It also intends to combat public health risks associated with poor water supplies, stop illicit activities sometimes conducted in these spaces, and halt the degradation of the landscape.

Nature-based Solutions characteristics

Currently, there are 21 horticultural parks across Lisbon, making up 850 plots across nine-point-five hectares. These parks include individual plots for cultivation, fences, storage cabinets, and irrigation water, and the use of pesticides is not allowed. They are created either on derelict land or in spaces with a previously high concentration of informal allotment gardens. There are two types of horticultural parks: "social parks", that focus on food production and self-sustainability (minimum area of 150 square meters), and "recreational parks", that mainly aim to contribute to users' overall life quality, by increasing their contact with nature and recreational opportunities (minimum area of 80 square meters). Both parks are inserted in urban parks and gardens, with other facilities (playgrounds, kiosks, etc), to enhance the multifunctionality of these areas.

Governance

The Urban Allotment Garden Programme was conceptualized under Lisbon's Green Plan, a manifest of values without legal value that included other programmes related to nature-based solutions for water efficiency, massive tree planting programmes, etc. This Programme was later included in the Municipal Ecological Structure. The horticultural parks are regulated by the municipality, with the number of plots in each park depending on available space. Each plot owner is required to pay an annual fee that covers the space to cultivate products and water for irrigation (particularly important during summer). The plots are attributed by public contest according to the proximity of the plot owner's house. This proximity-based criterion is intended to ensure local stewardship and reduce the risk of neglect. However, this rule may inadvertently limit access for residents in underserved or densely built-up areas where green space is limited. As a result, the distribution of allotment gardens within the city may not fully align with the needs of populations who would benefit most from access to them, such as lower-income communities or those without private outdoor space. Despite this, the municipality has taken steps to promote inclusivity and integration that include the placement of these plots within public urban parks which are open to the broader community. The city has also established clear rules and guidelines to ensure uniformity, and provides technical support, training, and community-building activities to support urban gardeners. These efforts help to enhance the social and environmental value of the programme, though further consideration of equity in plot allocation could strengthen its impact.

Outcome

The multifunctional municipal horticultural parks create conditions for planning mixed uses in public spaces, with social and economic advantages for the population, increasing, at the same time, biodiversity, and ecosystem services, such as flood regulation and pollination. Plot owners often feel a strong sense of place to their plot, serving as an extension of their household. In addition, these spaces allow the older population to stay active and allow younger generations to learn about food production and increase their contact with nature.

Writing: Filipa Grilo
Review: Veronica Olivotto
Copy Editing: Veronica Olivotto
Design: Phuong Anh Bui

Recreational plots within a municipal horticultural park in Lisbon. This photo highlights the strong sense of attachment of the plot's owner to the space, depicting a crib that the gardener installed to accommodate their grandchild when visiting

References:

1. Grilo, F. et al. (2016) 'Allotment gardens in a historical city: perceived benefits and challenges', in. Growing in Cities: Interdisciplinary Perspectives on Urban Gardening, Basel, Switzerland, pp. 137–148.
2. PORDATA (2023) Estatísticas, PORDATA. Available at: https://www.pordata.pt/pt/estatisticas (Accessed: 21 July 2025).

Photo by Filipa Grilo.

EU_19 Finale Emilia Food Forest Supported by the Rumon Association

Context Overview

LOCATION

NBS TYPOLOGY	ECOSYSTEM TYPE
Restoration	Agriculture & food spaces; Green spaces (gen.)
IMPACT SCALE	**URBAN TYPOLOGY**
Site	Peri-urban
SOCIAL CHALLENGES ADDRESSED	**CO-BENEFITS**
Biodiversity loss; Social inclusion	Green jobs and economic benefits

Finale Emilia is a small town in northern Italy, located at the far eastern edge of Modena province, near the border with Ferrara. It lies in the lowland area known as the Bassa Padana and has about 15,000 residents (ISTAT, 2023). The town has a long history, dating back to the 11th century, but it still shows signs of damage from the 2012 earthquakes that hit the region. This part of the Po Valley, which stretches towards the Adriatic Sea, is known for having some of the worst air pollution in Europe (Giuffrida, 2023). The pollution mainly comes from heavy industry and intensive farming, both of which are key parts of the local and national economy. In the Po Valley, common crops include wheat and corn, and there is a high concentration of livestock farming, especially pigs and cattle, as well as many manufacturing industries.

Introduction to the case study

Rumon is an association for the promotion of social wellbeing established in early 2022 with the objective of reforesting the Bassa Modenese. It began its efforts through the creation of a food forest. Rumon collaborates with Manitese, a non-governmental organization that coordinates international cooperation projects to develop autonomous and sustainable economies in collaboration with local communities. The Finale Emilia branch, where Rumon operates, is one of the largest in Italy, with a volunteer force of more than seventy individuals. Manitese's rear entrance, overlooking the land utilized by Rumon, is open to the public and situated along a cycling path frequently used by residents. The land that Manitese entrusted to Rumon was originally farmland and it extends for 5,000 square meters, of which 3,000 already had plantations and 2,000 were uncultivated lawns with a small part (of about 400 square-meters) to become a vegetable garden area. Rumon originated from the initiative of four founders, then extended to 30 volunteers, with the initial goal of planting as many trees as possible, and soon evolved to embrace principles of biodiversity and social cohesion. A music festival catalyzed members of the public, disabled youth from the Ancora Center in San Felice sul Panaro where hobbyists, ecologists, experts, and musicians collectively planted the first 70 trees.

Nature-based Solutions characteristics

The idea behind creating a food forest was to build a space that encourages connection and cooperation, a welcoming place where both people and animals can find food and benefit from nature. Since November 2022, 200 trees and 200 shrubs have been planted. A key strength of the food forest is biodiversity. With the help of an agronomist, different varieties of plants were planted, taking into account the synergy between many species, following the principles of regenerative agriculture. The central part has an orchard of pome and stone fruits, which are ancient varieties from the Emilia-Romagna region. In other areas trees, such as hazelnuts, vines, pomegranates and mulberries were planted At the entrance to the vegetable garden there are shrubs with brambles, aromatic and melliferous plants (e.g. lavender and rosemary). Rumon is now also in charge of managing Manitese's greenery, including the vegetable garden. Every week several volunteers, asylum seekers and individuals in need from social services come together to manage the forest and vegetable garden. In spring 2024, the responsibility of caring for poultry, which are capable of consuming tall grass and fertilizing the soil, has been added to the Manitese's tasks.

Governance

Initially, Rumon was funded by its founders, who organized fundraising events, collected private donations to reach the goal of 4,000 EUR (4,212 USD). Manitese donated EUR 800 (USD 842) to reach this goal. The first larger financial help came from a local bank foundation (Fondazione Cassa di Risparmio of Mirandola), which donated 4,000 EUR (4,212EUR). This initial investment of 8,000 EUR (8,464 USD) made it possible to buy the irrigation system and to support the expenses of the first events. The project also relies on donations from private citizens. The music festival 'Free Trees' (Alberi Liberi), which allowed the first 70 trees to be planted, was made possible thanks to the regional call for proposals 'Roots for the Future' (Radici per il Futuro), which allows Emilia Romagna' residents to take advantage of a large number of young trees free of charge by booking them and picking them up at local nurseries. A second edition of the festival was organized in April 2023 to plant more trees. Starting from spring 2024, Rumon officially became a part of Manitese, gaining the benefits of belonging to an organization registered with the National Third Sector Register (RUNTS). Organizations registered there have state recognition and increased possibilities for funding. Since its registration in the RUNTS, the founder of Rumon became an employee of Manitese, but the project continues to rely on the strength of volunteers.

Outcome

The strength of this project is that it combines care for the environment with building strong relationships between people. In collaboration with Morando Morandi High School in Finale Emilia, aromatic plants were planted in order to create educational experiences for youths. The presence of a stable group of workers is assured thanks to the establishment of three collaborations: a local social cooperative (Cooperativa Sociale Aliante di Mirandola) establishes school-to-work training programs to combat school dropout; a cooperative managing the Reception Center for Asylum Seekers in Finale Emilia (CEIS ARTE) provides projects for integration; the social services of a close town (San Felice) support job placement.

The main challenges encountered include: 1) variable income, as it is difficult to predict the production of the food forest and the resulting products; 2) the ongoing search for external funding; 3) the substantial reliance on volunteerism, involving significant time investment in planning, meetings, and physical labor. The aspiration is to further expand this model into additional territories, develop outdoor exhibitions and cultural events, conduct workshops in schools and in the food forest for the benefits of youths and adults in need. Additionally, there are plans to establish a distillery for producing essential oils from aromatic plants.

A mechanical system to protect plants from snails, employing a net as a barrier instead of chemical poison

Transplanting with a compost bed, cardboard mulch, and drip irrigation for water-efficient plant growth

New animals at the food forest: donkeys to help compost production

Writing: Eugenia Castellazzi
Review: Veronica Olivotto
Copy Editing: Veronica Olivotto
Design: Phuong Anh Bui

Acknowledgements:

Thanks to Antonio Luzio and Elisa Debbi (president and vice president) for the interview, data and pictures provided.

References:

1. Giuffrida, A. (2023) '"Impossible to live like this": Italy's Po Valley blighted by air pollution among worst in Europe', The Guardian, 21 September. Available at: https://www.theguardian.com/world/2023/sep/21/italy-po-valley-blighted-air-pollution-worst-europe (Accessed: 17 December 2024).

2. ISTAT (2023) Bilancio demografico mensile, demografia in cifre. Available at: https://demo.istat.it/app/?l=it&a=2023&i=D7B (Accessed: 17 December 2024).

3. Rumon (no date) Rumon. Available at: https://rumonforeste.it (Accessed: 17 December 2024).

4. @rumonforeste (no date) Instagram. Available at: https://www.instagram.com/rumonforeste/ (Accessed: 17 December 2024).

Photos by Antonio Luzio and Elisa Debbi.

EU_20 Intercommunal Dune Management in Brittany

Context Overview

LOCATION

NBS TYPOLOGY Restoration	ECOSYSTEM TYPE Coastal (gen.)
IMPACT SCALE Greater urban region	URBAN TYPOLOGY Outside the city with urban impacts
SOCIAL CHALLENGES ADDRESSED Land degradation; Flooding	CO-BENEFITS Increased biodiversity and their functions

The Dunes of Gâvres ecosystem is located in the region of Brittany in France, which is also home to several natural areas listed as part of the European Union's "Natura 2000" network of sites important for protection. The dunes are adjacent to the commune of Gâvres (Fig. 1) and belong to a 25 kilometers long coastal ecosystem that hosts nearly 700 plant species, of which 80 are protected.

Introduction to the case study

The Dunes of Gâvres are a unique and inspirational example of coastal management practices which has led to beach regeneration including accumulation of sand that in turns is supporting plant species recovery and improved biodiversity. Important to the success of this region is the management of the dune ecosystem by an intermunicipal cooperation of 25 small communes, the Lorient Agglomération[4].

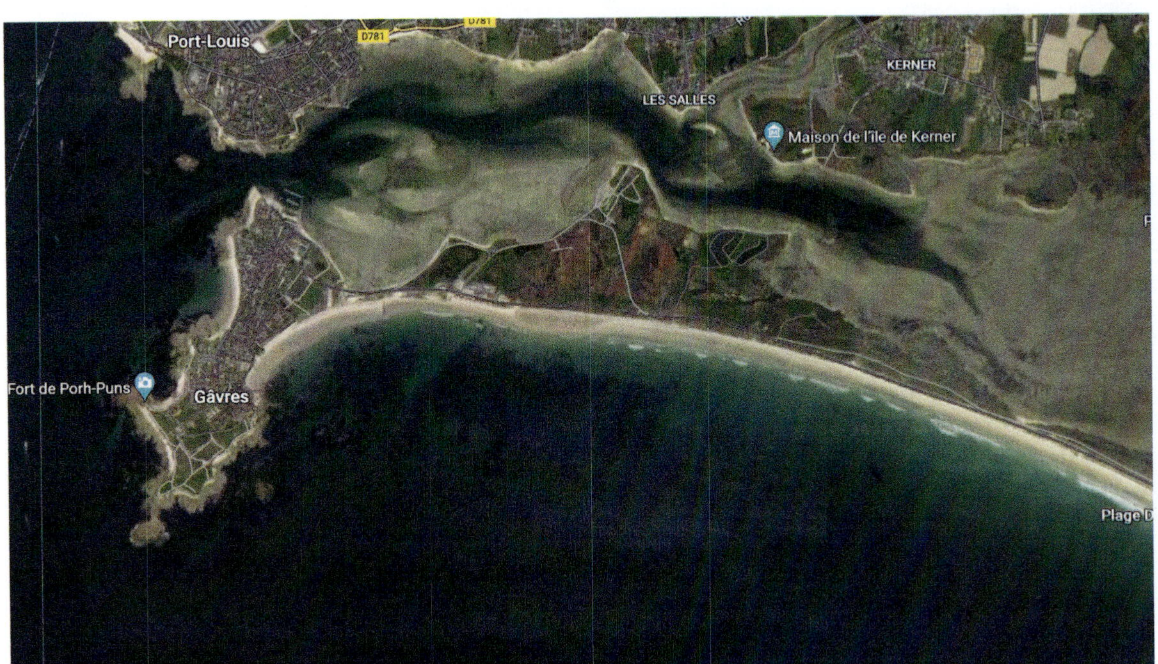

Map of Gâvres, the tombolo and the dune ecosystem in South Brittany, France located on a peninsula and connected to the mainland through a tombolo (a bar of sand). Taken from Google Maps.

While the year-round population in Gâvres is below 700 inhabitants, about one-point-five million people visit Gâvres every year for the beautiful beaches, especially during the summer season. Tombolos, the sand bar connecting the dunes to the mainland, are dynamic landforms that can naturally disappear and reappear over time, but urbanization, tourism and massive sand extraction have destabilized the tombolo that maintains the connection between Gâvres and the mainland. Shifts in sediment transportation and depositional patterns, together with sea-level rise, storms, and erosion caused by climate change pose important risks to the dune ecosystem and its potential as a nature-based solution for coastal biodiversity that also provides other ecosystem services.

In 2008, following concerns from a recent coastal storm, the Lorient Agglomération increased its coastline protection and dune management efforts to improve protection from future climate impacts. In 2015, the cooperative initiated a partnership with the University of South Brittany to conduct morphological and hydro-sedimentary monitoring of the beach. Framed as an nature-based solutions strategy for climate adaptation (*Solutions d'Adaptation Fondées sur la Nature or "SafN"*), Lorient Agglomération's strategy is to restore and reinforce the dunes to reverse erosion impacts and optimize the buffer effect of the dune ecosystem while preserving its fragile biodiversity.

Nature-based Solutions characteristics

Lorient Agglomération's dune management plan is focused on Gâvres' dune area and intertidal beach zone. The project, started in 2018, focuses on four main interventions:

1. Re-optimize existing hydraulic groins (wooden or rock barriers built perpendicularly to the shore) that limit sediment transportation
2. Construct three new wooden groins that allow the homogenous accumulation of sediments around the groins
3. Restore a degraded dune zone through a hybrid approach to re-elevate the dune with a sub-structure with kaolin and sand, protected with gabions, and on top of which beach grass (ammophilia arenaria) is planted for sand stabilization (a similar approach had been previously adopted on the nearby site of Larmor-Plage).
4. Regenerate embryo dunes with AlgoBoxes. Algoboxes are rectangular fenced areas in which algaes are trapped during high tides and help capture sand and propagate beachgrass This approach had also been previously tested in the region (e.g., the Rhuys peninsula)

Pre-existing groin

New wooden groins

[4] Communes are the local administrative units in France. They are analogous to civil townships and can vary in size and area.

Re-elevated dune

Algoboxes

Governance

The governance structure is important to the dune restoration and ongoing management. The Lorient Agglomération is the public administration body in charge of restoring the Dunes of Gâvres as per the GEMAPI national law which requires the joint management of aquatic ecosystems. Agglomérations are made of elected officials from the member communes to facilitate coherence in territorial planning and decision-making for transboundary resource management, and that allow the grouping of resources. About 500 000 EUR (about 550,000 USD) have been allocated to the project as per the end of 2023. However, officials estimate that twice this amount is needed to continue the strategy over the next few years.

The team of Lorient Agglomération continues to work closely with the University of South Brittany and with the mayor's office of Gâvres to align the project with wider urban planning considerations. Public engagement aspects of the project have included town halls and consultation with affected households before the start of the interventions. Negotiations were needed regarding the level and height of restoration of the dunes and limited passageways. Limiting touristic activities to make space for the dunes has been a contentious decision given that tourism is essential to the local economy. The nature-based solutions strategy has also been challenged by actors preferring relying on grey infrastructure to protect the coast from storms and erosion.

Progress and results have been assessed through a citizen monitoring program. A team of volunteers (the Coastal Citizen Observatory) regularly monitors the level of sand on the groins using a protocol developed by the University of Southern Brittany. Besides, any visitor can take pictures of the dune and upload them on an online platform (Coast Snap app). Coast Snap helps collect images that are collated in a database managed by the university. Monitoring is increased from October to April during which storms are more frequent.

Outcome

The project is widely viewed as a bright spot in nature conservation and protection to ensure the dune ecosystem can serve as a nature-based solution over time. The project has received the Life ARTISAN national prize for its successful initiative to adapt to climate change. Key outcomes include sand accumulation of up to 1 meter observed within less than a year after the beginning of the interventions on some parts of the beach. The progressive accumulation of sand volumes has supported the rapid expansion of biodiversity. Reintroduced plants include beach grass, wallflower, and sea holly. Many emblematic bird species for this dune ecosystem, including endangered species like the Kentish plover, are recovering.

Writing: Loan Diep
Review: Natalie Pierson
Copy Editing: Timon McPhearson
Design: Phuong Anh Bui

Acknowledgements:
Julien Thébaud

References:

1. Grand site de France (2021) Dunes Sauvages de Gâvres à Quiberon, Réseau des Grands Sites de France. Available at: https://www.grandsitedefrance.com/component/content/article/18-les-sites-membres/80-dunes-sauvages-de-gavres-a-quiberon.html (Accessed: December 17, 2024).

2. Sedrati, M. (2018) "A sustainable alternative for coastal dune restoration by sand-trapping fences and algae wrack: AlgoBox ®," in. EGU General Assembly 2018, Vienna, Austria.

3. Sedrati, M. et al. (2025) "Nature-based solutions for coastal dune restoration: The case study of AlgoBox in South Brittany, France," Ecological Engineering, 210, p. 107440. Available at: https://doi.org/10.1016/j.ecoleng.2024.107440.

4. Sedrati, M. and Cochet, J. (2015) "Un outil écologique pour la régénération des pieds de dunes grâce aux échouages de macro-algues," in. Conférence Méditerranéenne Côtière et Maritime, Ferrara, Italie, p. 4. Available at: https://doi.org/10.5150/cmcm.2015.017.

5. Trophées de l'Adaptation au Changement Climatique Life ARTISAN (2021) Aménagement Dunaire AU Service de la Lutte Contre la Submersion Marine. Trophées de l'Adaptation au Changement Climatique Life.

Photos by Julien Thébaud and Loan Diep.

EU_21 Green corridor alongside the Südpanke stream

Context Overview

NBS TYPOLOGY	ECOSYSTEM TYPE
Restoration	Rivers & waterways; Forests & trees
IMPACT SCALE	**URBAN TYPOLOGY**
Site	Urban core
SOCIAL CHALLENGES ADDRESSED	**CO-BENEFITS**
Physical and mental health; Nature connection	Place regeneration & urban revitalization; Recreational benefits; Human health and well-being; Increased biodiversity and their functions

Berlin is a dynamically growing city with a population of three-point-six million. It is located in Central Europe at the junction of maritime and continental climates and receives about 520 millimeters of annual rainfall, with significant seasonal and annual variations. The marshland in which the city was founded was formed by the drainage system of several glacial periods. Berlin remained a collection of villages until around 1800, when population growth and the Industrial Revolution dramatically influenced and changed the rivers, marshes and built structures in Berlin. As the city grew, canals were built, some rivers were channeled underground and the groundwater table changed dramatically—creating sustainability issues for the city today.

Introduction to the case study

The green corridor alongside the Südpanke stream is an example of how the city of Berlin increased its efforts to develop green infrastructure to meet the European Water Framework Directive and adapt to climate change while increasing local biodiversity potential. The city also pays attention to environmental justice: the Umweltgerechtigkeitsatlas Berlin (Environmental Justice Atlas of Berlin) is a policy tool developed by the Berlin Senate to map and analyze the intersection of environmental burdens and social disadvantage across the city. Major goals besides those are increasing the landscape connectivity both for wildlife as well as for pedestrians and bikers (Wellmann & Haase, 2023).

Nature-based Solutions characteristics

On the site, built and ecological infrastructure are well mixed and coordinated. Built infrastructure such as playground equipment and benches are developed around already established spontaneous vegetation. One side of the stream is built up and suitable for pedestrians and cyclists, while the other side offers a gradual slope, making the water more accessible to wildlife. Biodiversity promoting measures such as wildflower meadows are intermixed with more regularly cut lawn that is more accessible to park users. These characteristics support social interaction while promoting local biodiversity.

Governance

Construction was led by the GrünBerlin foundation (landscape construction) and the Berlin Senate for the Environment (water construction). The state of Berlin and compensation payments from construction sites funded a total of 9.87 million EUR (13 million USD) of which 6.95 million EUR (9.2 million USD) was spent on the construction of the water features. Most of the landscape-related construction was funded from other mitigation measures. Construction started in 2010 and was completed in 2021 (GrünBerlin, 2021).

Outcome

There are challenges in terms of water quality and connectivity due to the location and history of the site. Water levels and water quality can vary due to changing rainfall patterns. Water quality issues are related to the fact that the Panke River used to run alongside heavy and polluting industry upstream, which still has a negative impact on water quality and makes the interaction with the water only partially safe. In addition, downstream connectivity is still lacking, as the river turns into pipes after the newly constructed section, limiting the potential for aquatic organisms to migrate upstream. In general, however, the intervention—although comparatively expensive due to the extensive overhaul of the water infrastructure—can be seen as having a very positive impact on the surrounding area, improving walkability, cooling and biodiversity (von Hagen, 2021).

The river Südpanke as seen from Ida-von-Armin Street in central Berlin

Newly created park space and its maintenance during a heat wave

Information sign promoting floral diversity for higher pollinator diversity on the site. In the back is the main building of the German intelligence service

A newly installed playground installed into previously existing spontaneous vegetation

Writing: Thilo Wellmann
Review: Veronica Olivotto
Copy Editing: Veronica Olivotto
Design: Phuong Anh Bui

References:

1. GrünBerlin (2021) Factsheet Grünzug Südpanke, 2 Bauabschnitt. Available at: https://gruen-berlin.de/fileadmin/user_upload/Downloads/gruenzug_suedpanke/gb_projekte_gruenzug_an_der_suedpanke_faktenblatt.pdf.

2. von Hagen, J. (2021) "Berlin Südpanke offengelegt," Garten + Landschaft, 16 June. Available at: https://www.garten-landschaft.de/berlin-suedpanke/ (Accessed: December 18, 2024).

3. Wellmann, T. and Haase, D. (2023) "Blue Nature-based Solutions in and of Cities," Bringing Nature Back, 16 March. Available at: https://bringingnatureback.com/research-news/research-stories/blue-nature-based-solutions-in-and-of-cities/(Accessed: December 18, 2024).

Photos by Thilo Wellmann.

NATURA

Printed and bound by CPI Group (UK) Ltd, Croydon, CR0 4YY
18/09/2026
14965071-0001